Introductory Note

The contents of this volume consist of past and present material contributing to the beginning of a discussion within the Socialist Workers Party on the farm question. As stated at the end of the first item, the report by Doug Jenness on "American Agriculture and the Working Farmer" that was approved by the National Committee of the SWP in plenary session, May 1979, "It should be understood that this report is only the beginning of our discussion of the farm question. Its purpose is to begin the process of getting the party to think and learn more about this question. Hopefully this can lead to a written resolution at some point."

Today's interest and concern on the part of revolutionary socialists in the farm question reflects a rich and fruitful Marxist history. One example is the attitude and policy taken by Lenin, Trotsky, and the Bolsheviks and their ability to weld together an alliance of workers and peasants aimed toward getting rid of Tsarist oppression and capitalist exploitation and establishing the world's first workers state.

Another outstanding example is the policy toward the peasants followed by the revolutionary government in Cuba headed by Fidel Castro.

In addition to the rich theoretical and practical experience on this question reflected in the articles and speeches by Engels, Lenin, and Castro included in this volume, the American Trotskyists have also had some practical experience of their own. During the 1934 Teamsters strike in Minneapolis an alliance was forged between the striking workers and working farmers. Farrell Dobbs, a central leader of that strike, discusses this in *Teamster Rebellion* (the first of four volumes on the Midwest Teamsters in the 1930's):

"With another strike looming as a certainty, attention was turned to shoring up Local 574's alliances. An agreement was reached with three farm organizations: The Farmers' Holiday Association, the National Farm Bureau, and the Market Gardeners Association. It provided that the union pickets would not interfere with farm trucks during the strike if they carried permits from Local 574 and the farm organization to which each operator belonged. To prevent chiseling on this arrangement farmers committees undertook to picket roads leading into Minneapolis. When put into practice later on, the procedure worked well. Difficulties experienced during the May action were largely avoided, and the union enjoyed general sympathy among the farmers. Sentiment favorable to the union was further enhanced by the fact that the dual character of the permit system enabled the farm organizations to conduct effective recruitment drives of their own. . . . The system proved to be so successful that it was continued even after the next strike was over and appreciation was shown to the union by donations to its commissary."

Dobbs explains that this policy of a striving for an alliance between farmers and workers was not invented by the Trotskyist leaders in Minneapolis. They were applying essentials learned from past Marxist leaders. Earlier efforts to unite working farmers and proletarians into a political alliance in the U.S. ". . . had the approval of the Third International, then led by V.I. Lenin and Leon Trotsky."

Some skeptics who may not think the question of farmers to be important today might argue that conditions in America are certainly unlike those that existed in Russia where a majority of the population were peasants; and they are unlike the 1930s when there were many small farmers, and many sharecroppers for that matter.

Weighed against these doubts are the important

recent protests of farmers, most notably the formation of the American Agriculture Movement and the wave of tractorcades involving thousands of farmers. Farmers were also active in aiding striking coal miners in early 1978, and in their support to the successful campaign to defeat the "right-to-work" referendum in Missouri.

These events, occurring at a time when the Socialist Workers Party was implementing its turn into basic industry, helped us see the importance of the farm question to the struggle against capitalist rule. And they have engendered the opening of an educational discussion within the SWP designed to identify, assemble, and evaluate new facts and changes that have taken place in American agriculture.

During the course of this discussion, a broad range of questions have been raised: "What is 'parity'? How does it work? Why is this a main demand of the farmers?" Other questions revolved around an assessment of the social weight of farmers today as compared to their numbers and their importance to the working class and the unions.

The need for answers to these and similar questions stimulated the beginning of a fruitful study within the SWP. Members of the SWP and supporters attending the 1978 Active Workers and Socialist Educational Conference in Oberlin, Ohio, participated in two classes presented by John Staggs and Syd Stapleton. They centered on a study of American agriculture, its facts, figures, history and political significance for revolutionary socialists.

Alongside this, the *Militant* newspaper sent one of its reporters, Osborne Hart, to different areas and featured frequent coverage of emerging struggles launched by working farmers throughout the country. *Militant* readers were able to gain firsthand information on these events.

And participating in this discussion were members of the SWP who reflected earlier experiences as farmers or farm workers.

The fruits of this collective effort were represented in the report by Doug Jenness listed in this volume.

The second item in this volume is a contribution by one of the founders of scientific socialism, Frederick Engels, written in 1894 entitled, "The Peasant Question in France and Germany." Engels discusses farmers in two imperialist countries at that time. His opposition to a policy of forced collectivization warrants study in light of the policy Stalin followed in the Soviet Union.

The third, "Theses on the Agrarian Question" was drafted by Lenin and adopted at the Second Congress of the Communist International (Comintern) in 1920.

The fourth section is a report presented by Lenin on work in the countryside to the eighth congress of the Russian Communist Party, March 23, 1919.

The last two items are speeches by Fidel Castro on agriculture in Cuba: "Cuba's Agrarian Reform," given August 18, 1962 to the National Congress of Cane Cooperatives; and the talk given to the Third National Congress of the ANAP (Asociación Nacional de Agricultores Pequeños—National Association of Small Farmers) May 18, 1967.

Additional readings are recommended by Doug Jenness in the summary of his report.

Paul Montauk

AMERICAN AGRICULTURE AND THE WORKING FARMER
By Doug Jenness

One result of the party's turn into industry is that we're discovering firsthand how U.S. industry and agriculture really works. We're learning how goods and services are produced in our society, who produces them, and who distributes them. We're discovering who our allies and friends are, those who should be treated as fellow workers, and those who are our enemies.

This has placed the question of agriculture and working farmers on our agenda for the first time in several decades and induced us to take a fresh look at this sector.

Our discussion is important because we've slipped a bit on the farm question. Let me give you an example.

In George Novack's introduction to *The Transitional Program for the Socialist Revolution*, by Leon Trotsky we say:

"We haven't worked out a program for the American peasants, because we don't have so many. There are only about 5 million farmers compared with 9.5 million college students in this country. Thus the student population, which is in motion, is more important than the farm population, politically speaking. Indeed several years ago we amended our transitional demand for a new regime from 'a workers and farmers' government' to 'a workers' government.' In some areas—for example, in Minnesota—we may require an extensive set of demands for the small farmers in the future as we have had them in the past. However, this is not now a central concern for us on a national scale."

I should say at the outset that this paragraph has appeared in all three editions of the book, which many of us have gone over closely. But now, as we come to grips with the farm question again, we can recognize that the approach is wrong. I raise this example, not to denigrate the struggles of students, which were exemplary in the 1960s and from which many of us were recruited. The point is to indicate how we had begun slipping into underestimating the social weight of working farmers in the struggle to overturn capitalism and consolidate a workers state, and the decisive importance of this question for the labor movement.

Part of the reason is that the number of farmers has declined so much since World War II that there's a tendency to consider the farm question in the United States and other imperialist countries as a peripheral issue. It's really only a question, it's believed, for the semicolonial countries like India, Iran, or Mexico.

This is not true. It's not a minor nor peripheral question, but a central one for the class struggle. It's likewise a central question for the Fourth International and its sections in imperialist countries, from New Zealand and Australia to France and Germany. In fact, the New Zealand comrades are ahead of us. They discussed this question at their last national committee plenum and have been writing some excellent articles in their press.

Importance of agriculture

A clear understanding of the role of agriculture is not an optional matter for the working class, but essential for its struggle for power. In the United States agriculture is the largest industry, employing some 20 million people directly or indirectly. Every one of us eats. This morning when you got up you ate something. And you put on some clothes. Where did that food and fiber come from?

Report adopted at the SWP National Committee Plenum, May 1, 1979.

Somebody had to cultivate the soil, plant the seeds, harvest the crops, and milk the cows. Somebody had to transport it to the flour mill, to a dairy, to a stockyard, or to a textile plant. Somebody had to process it, package it, and transport it to a supermarket, a restaurant, or a clothing store where it's sold.

We are all totally dependent on this industry. There are not enough backyards in this country to grow food to feed everybody. And even the option of going back to the farm in bad times, as millions did during the depression in the 1930s, is not realistic today.

Agriculture is also a major factor in U.S. foreign trade and consequently in U.S. foreign policy. It accounts for 20 to 25 percent of all U.S. exports. U.S. grain merchants export more wheat and corn than the rest of the world combined. It's one of the few economic sectors today where productivity is so great that the U.S. has a big edge over its competitors in other countries. One American farmer can produce at least 3 times as much as his counterpart in Western Europe.

Working farmers

How is the production of food and fiber organized? Who owns the land? Who does the work? Who profits from the tremendous productivity of American agriculture?

The first thing to understand about agricultural production is that farmers aren't a single class but a set of classes, in which both exploiters and exploited are included.

Among the exploited are family farmers, that is, farmers who utilize the labor of family members rather than hired labor. In the West many family livestock producers prefer to be called "ranchers" and use the term "farmer" to designate producers who till the soil. For the purposes of this report, family farmers will include *all* agricultural producers who use primarily family labor.

One common misconception is that family farmers have been virtually wiped out and replaced by industrial factory-type farms. Many people assume that agriculture has become more and more organized like manufacturing, with big capitalist enterprises hiring lots of wage labor, and that farmers utilizing the labor of the family are simply irrelevant relics of a past age.

It's true that the number of farms has been drastically reduced, that more than a thousand are wiped out each week, and that the number of farms is a little more than 2 million—about one-third of what it was in 1940. At the same time, the average size of these farms has considerably increased. *But the fact remains that family farms account for over half of farm output.*

But that doesn't tell the whole story. Most of certain key commodities such as wheat, corn, soy beans, milk, pork, and much of the beef are produced by family farms. Even a large proportion of chickens and eggs are produced by family farmers.

Semiproletarian farmers

About two-thirds of the farmers in this country get more than half their income from nonfarm sources, usually an outside job in a mine, factory, packinghouse, railroad or whatever. Most of them are family farmers. Some are just not making it and are really on their way out of farming. Others are hoping that the outside job is temporary and they can get back on their feet again. Some may have as much as several million dollars worth of land and machinery.

But all have to take a job because they're over their heads in debt. Jobs related to agriculture, especially ones located in farm areas, often employ many farmers or ex-farmers. This includes farm equipment plants, meatpacking, truck drivers, railroad workers, and so on. There are some industries, notably electrical and auto, that deliberately build plants in rural areas to take advantage of this type of labor supply. Other farmers commute, sometimes long distances, into cities to work.

We received a report last week from a comrade in our fraction in the large General Electric plant in Louisville, who described the composition of the workforce there. She listed the number of women, Blacks, and youth. And then she indicated—the first time I've seen this category in one of our union reports—that 10 to 15 percent of the workers are farmers or long-distance commuters from rural areas. It also turns out that the president of the IUE local in this plant is a farmer. As we get more comrades into industry we'll run across more and more of this.

This layer of semiproletarians or farmer-workers

is especially important in transmitting the problems of farmers to whole sections of the working class as well as helping to make farmers more sympathetic to the problems of wage workers. This was a big factor in the support that many farmers in Kentucky last year gave to the coal miners' strike and the big vote farmers in Missouri cast against the right-to-work referendum there last fall.

For example, comrades in Kansas City report that many of their co-workers work on the farm during the day and work second shift in the plants. It was these farmer-workers that were largely responsible for the outreach campaign to farmers during the fight to defeat the right-to-work referendum.

Small capitalist farmers

There is another category of farmers who see themselves as family farmers and use the labor of the family, but also hire wage labor. It's common for many family-owned vegetable and fruit farms to hire crews of migrant workers sometimes for only a few weeks. For example, in Minnesota the sugar beet growers hire migrant workers for just one part of the production process—weeding. The rest of production is mechanized enough so that it can be handled by family labor.

This layer of small capitalist farmers exploit labor and are hostile to the efforts of farm workers to fight for their rights and a decent living, but at the same time are exploited by the banks, processors, and distributors. They are in a sense middle farmers—in between the toiling independent producers who don't use wage labor and the big capitalist farms that regularly employ wage labor and are often owned and operated by big monopoly corporations. Furthermore, they aren't a homogeneous category. The smaller ones, who hire small amounts of wage labor for short periods and depend primarily on family labor, have many points in common with working farmers.

Big capitalist farmers

While many family farmers and small capitalist farmers are involved in the production of fruits, vegetables, and nuts, capitalist farms, owned and operated by large corporations, account for a large proportion of these commodities, especially in the Southwest, California, and Florida. Large monopolies like Tenneco, Del Monte, and so on, operate seed-to-supermarket-type operations. They own the land, contract for the labor they need, and directly process and sell the produce themselves. Some giants like Tenneco also own their own farm equipment, fertilizer, and pesticide companies.

Agricultural wage workers

Farm workers who work for capitalist farmers are an integral part of the working class. They are proletarian; and, as such, the labor movement should embrace them and support their struggles as fellow workers. Because of the seasonal character of the work, the capitalist growers have been successful in preventing the effective organization of most farm workers. The struggle of the United Farmworkers Union over the past decade was the most successful effort to date to organize farm workers. It became a major social issue with broad support throughout the country. And it succeeded in winning some contracts with grape and lettuce growers. But the failure of the AFL-CIO leadership to mobilize behind the farmworkers meant that their gains were limited. Their prospects remain in jeopardy as the growers mount an offensive against this new and weak union.

The absence of adequate union organization through which farm workers can defend themselves has meant that they suffer abysmal conditions. Wages are low while health and safety conditions are atrocious. There are no unemployment or health benefits. They have little protection against the arbitrary and sometimes savage treatment of the overseers hired by the growers to supervise work. And they are made the victims of increased productivity resulting from the greater mechanization of agricultural production. The worst conditions are suffered by the migrant workers, who move from one part of the country to another following the harvests.

There are about three million agricultural wage workers, according to government statistics, including hired hands. The figure may well be higher because not all the employers accurately report the number of employees. Only about one-third of these workers are full-time. The majority of farm wage workers are white; but Chicanos, Puerto Ricans, Blacks, and Filipinos make up a disproportionate amount of those who are migrant laborers. It's interesting that this figure hasn't changed much over the last several decades. This destroys another misconception—that as family farms are wiped out, more and more wage labor is used.

Why the family farm continues to exist

Why hasn't all of agriculture gone over to the industrial form of organization? Why don't the big monopolies and the banks just take over the wheat, corn, and soybean fields and the dairy herds, establish massive farms, and hire wage labor to work them?

One reason is that under the present setup they can get the working farm family to take all the risks—the burden of bad weather, unstable market conditions, high interest, and taxes. Furthermore, members of the farm family who labor on the farm aren't paid by the hour. Because the farm is theirs, they put out a tremendous amount of work, fourteen, sixteen hours a day if necessary, during planting or harvest time. As property owners and as owners of the product they produce, they feel responsible for the farm—its machinery and land—and for organizing production. This fact is skillfully used by the capitalist rulers to try to get the farmers to see themselves as businessmen, like themselves; as something different from labor and opposed to labor; and as a conservative and "responsible" force in society.

Not only is monopoly capital aware of the advantages of getting the independent producers to take the risks, they also recognize that family farmers are less likely than wage workers to organize collectively against them. They can put the squeeze on working farmers and face less risk of strikes, union organization, etc. The wage worker, who sells his labor power, doesn't feel that he owns, controls, or has any rights regarding the product he produces. But the working farmer does feel that he owns the product, controls it, and has the right to market it. He is therefore much more concerned if it is destroyed as the result of protests against his exploiters than is the wage worker.

Take the situation with tomato farmers in Ohio. For many years the growers there have hired migrant workers for a month or so in the late summer and early fall to harvest their tomatoes. During the past decade farm workers have been organizing.

Last year the Farm Labor Organizing Committee (FLOC) called a strike against Campbell and Libby, two giant food processors, demanding that they pay more to the growers so they could pay more to the farm workers. This approach was designed to win sympathy from the growers who are exploited by the food processors.

But Campbell and Libby wouldn't bargain with the farm workers on the grounds that they are employed by the growers, not by them. And the growers, who directly exploit the farm workers, not the processors, were very hostile to a strike at harvest time. Consequently, the farm workers won no contract.

In response to this threat the monopoly processors are prodding the growers into going over to big, self-propelled mechanical tomato pickers that won't require "unreliable" wage labor. Campbell Soup Company, for example, refused to contract this year with growers who do not use mechanical harvesters. Libby is insisting that a portion of the tomatoes it uses be mechanically harvested.

In 1975, 10.4 percent of the Ohio tomato crop was harvested by machine; in 1978, 29 percent; and in three to five years, it will be virtually the entire crop. Since the machines cost $40,000 each, some of the smaller, less competitive farms are being wiped out and the bigger ones are becoming bigger.

This process is occurring throughout the country and is already completed with many other crops, especially in California, where it is a big issue.

Mechanization has actually limited the growth of the agricultural labor force and in some cases, like the Ohio tomato growers, has led small capitalist farmers to go back to relying simply on the labor of their family.

How working farmers are exploited

Most working farmers don't want to give up being farmers. They want to be their own boss and organize the labor of their own family. They want to choose what day to plant, what day to cultivate, how many acres of corn or soybeans to plant, whether to have Holsteins or Guernseys, when to market their produce, whether or not to buy or rent more land, whether to borrow money to purchase a new piece of equipment, or try a new technique. And those who rent their farms usually want to get enough money to buy their own farm.

But this freedom, this independence, is more illusory than real. Working farmers, in the words of Frederick Engels, are "debt slaves." They are not capitalists, not even small capitalists. They don't accumulate capital, hire labor, or realize profit. They're not exploiters, but exploited. With their labor, with the work of their hands, they create a product. They sell their product in the market, but only get a small portion of it back for their own account. The rest is expropriated from them—stolen from them—along the way by the banks and trusts.

The banks take their share through interest payments on loans. In order to buy a new piece of equipment—some tractors cost as much as $100,000 each—or a new piece of land (land prices are soaring), so they can compete, farmers have to get bank loans. In order to get money for operating expenses—seed for the next harvest, fertilizer, diesel fuel, pesticides, feed for the livestock—they must borrow against their next crop. They are always in debt. They never climb out of debt. They're always paying interest on that debt. They are like the figure in Greek mythology, Sisyphus, who was condemned by Zeus to roll a rock up a hill only to have it come rolling back down again time after time before reaching the summit.

Then the big corporations, like John Deere and International Harvester, take their cut by charging monopoly-rigged prices for farm machinery. The big feed companies, like Ralston-Purina, take theirs. The big chemical companies, like Dow and DuPont, that sell pesticides and herbicides, take theirs. Exxon and all the others in the energy trust that sell fuel for the tractors and other farm machines, take theirs. All the monopolies that sell farmers the things they need for production exploit them.

Squeezing the farmer at the other end of this vise are the big processing and merchandizing trusts, who rig the prices they pay farmers so as to keep them as low as possible. For example, two giant monopolies, Cargill and Continental, handle half of all grain exported from the United States. They operate grain "pipelines"—all the way from farmer to foreign consumer. They own shipping companies, grain elevators, communications systems, espionage networks, and processing plants. And they cloak their business dealings from the public with a blanket of secrecy.

They are what is sometimes called "the middlemen." Did you ever think about what a middleman is? It's one of those words designed to cover up, not explain. It evokes the image of a person in the middle with one hand reaching out to take some produce from the farmer and then passing it along to the consumer in the supermarket.

The real truth about the middleman is that it is a category that includes two opposing classes—workers and employers. When the produce leaves the farm it is driven by truckers—either owner-operators or employees of big trucking companies. It is taken to a packing house, a dairy, or a cannery, where other workers process and package it. Then drivers or rail workers carry it to supermarkets and restaurants, where other workers sell it in the retail trade. More value is added to the produce as it goes along the line.

There the same food monopolies, the same "jolly green giants," who exploit the families that grew the food, also exploit the wage workers who process, transport, and distribute it. There's an immediate common tie here between the exploited independent petty commodity producer and the exploited wage-paid processor, transporter, and distributor.

In a few weeks, the Amalgamated Meat Cutters and Butcher Workmen, which includes packinghouse workers, will merge with the Retail Clerks Union, which is the largest union in food retailing. This merger will establish the United Food and Commercial Workers Union, the largest union in the AFL-CIO and a potentially powerful organization of food processors. It will have a lot in common with working and exploited farmers and many of its members will be farmer-workers.

The exploitation of the working farmer can be summed up in the expression often heard in the protests that the American Agriculture Movement has been organizing for the past year and a half: "We buy retail, sell wholesale, and pay the freight both ways."

The result is that the costs of production keep soaring for the farmer while the prices they receive don't keep up with these costs. And in the last couple of years, it was this growing gap that sparked the militant tractorcades throughout the country. They were reacting to the worst situation facing whole sectors of the farm population since the 1930s.

The anarchy of the capitalist market and the stranglehold of the banks and trusts mean that the farmers have little control over their lives. And as the economic crisis deepens, and the shortages and price explosions increase all through the economy, the plight and insecurity of the farmer will worsen. The present monopoly-contrived oil shortage, which is propelling fuel prices upward, is a case in point.

In a few ways, working farmers are even worse off than some unionized industrial workers. Many farmers don't have any medical insurance or pensions. And there are many restrictions on how much, if any, social security farmers can get. They don't have cost-of-living clauses. And farmers that don't make the costs of production and can't get another job are sunk in poverty. Consequently, there are big pockets of poverty in rural areas in this country.

One noteworthy effect of driving farmers off the land is that the South has been transformed. Hundreds of thousands of Black and white farmers have moved into cities and become wage workers. Large numbers moved from the South to get industrial jobs in the North. In some places, like North Carolina and South Carolina, industry has moved into the rural areas to employ these workers. As workers move South today they go not to farm, but to get jobs in big or small towns.

Black farmers

This transformation has especially affected the Black population. At the end of World War II most Blacks in the United States lived in rural areas. Today the overwhelming majority are part of the wage work force. There was a deliberate drive, especially since World War II, to drive these farmers off the land.

One common mechanism for this racist drive is through an institution called Heirs Property. During Reconstruction many Blacks were able to buy property and received titles from the Freedman's Bureau. Most owners, when they died, didn't leave wills and the land passed by statute to the wife and children. When they died the land passed to the grandchildren. Two or three generations later 100 people could be part owners of 100 acres, even though only one family might be farming the land. The difficulty with this is that, unless the farmer can round-up signatures on "quitclaim" deeds from all the living heirs, he cannot sell the land or use it for collateral for housing or for production. The Federal Housing Administration won't even provide loans if there is a "cloud on the title."

These restrictions facilitate land developers and speculators to purchase fractional interests in the land from any one of the heirs and then demand that "his" interest be partitioned off. Often the land cannot be easily or equitably subdivided. The courts, which don't care a whit about Black farmers, will order a sale of the land and a division of the proceeds. It is not uncommon for the person triggering the sale to purchase the entire tract, that having been the plan.

Whereas there were about one million Black farm families in the South in the 1920s, today only 40,000 remain.

How farmers have fought back

The particular problems facing farmers today aren't new. Farmers have been fighting to alleviate them for more than a hundred years, through their own direct action and by establishing dissident political parties. These included the Grange movement and Populist movement in the last century and the Farm Holiday movement of the 1930s.

Out of the Grange movement cooperatives were established. These were organizations of farmers set up to collectively buy farm machinery, seed, fertilizer, etc., in large quantities. They were then sold to co-op members at a cheaper price than they would have paid if they had bought them individually from the manufacturer. Similarly, they pooled their efforts in selling their produce to try to get the best price. While these farmer co-ops still serve their original purpose in some instances, many have long since been taken over by the banks or the food trusts. They have put their representatives on the co-op boards, invested capital in them, and transformed them into capitalist processing and merchandizing firms. Rather than serving the farmers, they have institutionalized the exploitation of farmers.

The dairy industry is dominated by such "co-ops." For example, Land o' Lakes, based in Minneapolis, is one of the largest dairy distributors in the country. It began as a dairy producers co-op and is still formally a co-op to which farmers belong. But it is also one of the biggest food corporations in the country, investing capital in a wide range of enterprises. It has extensive laboratories for research and development.

The largest farm organization is the Farm Bureau, which grew out of the government-financed county agent extension service in the early part of this century. It didn't start as a protest movement or organization of farmers and in fact received a boost from the dominant business interests. Today it's a huge business empire including insurance, oil, fertilizer, and finance companies. It operates scores of other co-ops which are capitalist businesses. A large percentage of its members have nothing to do with farming, and over the years the farmer members have increasingly become simply customers. The producers have no decision-making power. They don't get any special prices or benefits from being members of the co-op. It's just a big business tied in with the banks.

Another attempt by farmers to fight back against their situation was to force the processors into collective-bargaining agreements for prices that would guarantee a living income. This was the approach of the National Farmers Organization which was established in the 1950s and organized some militant protests in the 1960s. Their tactic was to get farmers producing a particular commodity, like milk or hogs, to hold their produce off the market in order to pressure the processors to sign a better contract. They compared this tactic to workers withholding their labor power from the employers during a strike. Their slogan was for "the cost of production plus a reasonable profit." (They don't mean profit in the scientific sense of that word, but enough income above production costs so that their families are guaranteed a decent living.) They established picket lines to prevent scab produce from going to market, and there were pitched battles where several farmers got killed. But they were unsuccessful in getting any contracts.

Government subsidies

Many NFO members have also played leading roles in the more recent protests organized by the American Agriculture Movement. The approach of the AAM is to seek government support for farmers. Since the 1930s there have been government programs ostensibly for the purpose of aiding farmers, and the Department of Agriculture was established as an apparatus to implement these programs.

The whole range of programs, from price support formulas to taking land out of production, do not help, or only inadequately help, working and exploited farmers. They are designed to help the capitalist farmers, the profit-making farmers, and the food monopolies. Only a minority of the farmers in this country get any type of subsidy whatsoever from these programs. And these are the best-off farmers. They are "farmers" like President Carter, who raked in thousands and thousands of dollars from government subsidies for his family peanut business. These programs are riddled with corruption, including farmers getting subsidies on crops they did not plant, yet marketing another product grown on that same land.

At the present time all the subsidy programs are pegged at taking land out of production. The idea is to use government intervention to reduce production, creating less meat, wheat, or whatever on the market, and driving up prices. Agriculture, like manufacturing, has alternating gluts and shortages. When there's a glut of steel on the market, U.S. Steel, Republic, and the like lay off workers and cut back production.

In agriculture, when there are gluts and prices are falling, the government attempts to encourage cutbacks by giving subsidies to the farmers. This especially benefits the largest farmers with the most land and the most hired labor. It's easier for them to cut back, save some of their production costs, still keep a lot of land in production, and get big subsidies on the side. The family farm—providing its own labor, operating on tight margins, with loads of debts, and a limited amount of land—wants to get the full use of its machinery and land in order to get the biggest crop possible. For it to take land out of production doesn't mean it can lay off workers. The "soil bank"-type programs are less beneficial, or not beneficial at all, to the family farm. And of course the irrationality of cutting back food production, of not utilizing the full productive capacity of American agriculture for the benefit of the hungry here and elsewhere, is totally reactionary. This policy is not in the interests of working and exploited farmers, it's not in the interests of working people, and it's not in the interests of millions of starving toilers throughout the world.

What is parity?
What should farmers demand? The AAM is demanding a hundred percent of parity. What does this mean? All this means is that, as the costs of production go up, the farmers want the prices for their produce to go up proportionately. They want to be able to meet the costs of production and have a living income. Parity is simply a way of measuring the costs of production in relationship to prices. An index was established based on the years 1910 to 1914 when there was supposedly a favorable relationship between costs and prices. One hundred percent of parity means that if the costs of production today are 50 percent higher than they were in 1910 to 1914, then the prices the farmer gets today should also be 50 percent higher than they were then. Supposedly then farmers would be able to meet the costs of production plus have enough to live on.

This scheme is similar in concept to the government's consumer price index which is used to measure inflation and to calculate escalator clauses in union contracts. One of the flaws in using the 1910 to 1914 index is that it wasn't really such a favorable period for farmers. The Roosevelt administration which first used this index in the 1930s said this was a favorable period. But that was only relatively true. Furthermore, the factors of production are quite different than they were in 1910 to 1914. Tractors weren't used as much then, fuel oil wasn't a factor nor were pesticides. Many of the costs of production are quite different. So it's not an accurate guide to calculating the real relationship between the farmers' costs and prices, any more than the consumer price index is really an accurate guide for calculating cost of living increases for wage workers.

By demanding 100 percent of parity, farmers are asking the government to make up the difference between the costs of production and the market price they are getting. We support this struggle. We believe the government *should* guarantee full cost of production and a living income to working farmers. But we say that there should be no subsidies whatsoever for capitalist and profit-making farmers, and no subsidies that are pegged to curtailing production.

We also call on the government to provide interest-free credit for farmers and free medical insurance and retirement pensions.

Who will pay for this? The employers attempt to turn workers against farmers by accusing farmers of free-loading off the taxes workers pay to the government. That was the line of the *New York Times* during the tractorcade in Washington this winter: It asserted that the protesting farmers were seeking "their own kind of permanent welfare payments." Secretary of Agriculture Bergland accused them of being "greedy seekers of profit." But leaving aside the fact that it is far better for our taxes to be used to assist fellow toilers than to the bloated war budget, we don't believe this well-deserved assistance should be paid from the taxes of working people. It should come out of a big increase in the taxes of the energy trusts and the food monopolies. Exxon, Tenneco, Pillsbury, Cargill, and the other exploiters should be forced to give back some of the money they've robbed from working people—both farmers and wage workers—to help meet their needs.

But there is another demand that the AAM leadership has been raising which is not so good. This is the demand for import quotas and tariffs on farm products from other countries. Some products are produced cheaper in other countries such as lamb in New Zealand and tomatoes in Mexico. But quotas on agricultural imports are just as reactionary in effect as they are on any other product. And the demand for restrictions by farmers is just as wrong as the demand by United Steelworkers president Lloyd McBride for quotas on foreign produced steel. Unfortunately, this is one case where the farm leadership has clasped hands with the labor bureaucracy.

Last fall in Colorado, the Rocky Mountain Federation of Labor and leaders of AAM joined in a common project of printing up thousands of bumper stickers with the inscription, "Protect our heritage—Buy American." But this would mean higher food prices for working people in this country, and the food trusts, not the farmers, would gain from this. The burden is also placed on the New Zealand sheep farmer and the Mexican tomato picker, pitting American farmers against fellow toilers in other countries.

Who's responsible for high prices?
The demand by farmers for government help is also used by the ruling class to try to drive a wedge between working farmers and wage workers by charging that such subsidies will increase supermarket prices. They say: "If farmers get a better price for their produce, it automatically means food prices will go up."

But at the same time the employers say to farmers that the reason they charge so much for farm machinery, fertilizer, transportation, and so on is that workers are demanding higher wages. Milk drivers, farm equipment workers, meat packers, and railroad workers are demanding too much, they claim.

What really goes on? Look at the settlement of the milk drivers strike here in New York City this weekend. The milk drivers and the dairy workers, organized by the Teamsters, got a small wage settlement, far less than a cost of living increase. The dairies then immediately raised the price on a quart of milk two cents; and that's being blamed on the workers.

The necessity of countering the employers' attempts to pit workers and farmers against each other was pointed to in an interview Ralph McGee, the executive secretary of the Kansas State Federation of Labor, gave to the *Des Moines Sunday Register* last fall. He said:

"We have common aims. We want to save the family farm from conglomerate corporations that already control most natural resources in this country except food. We want to help the farmers cope with the five or six giant grain companies and with those companies that control so much of the food processing business.

"And we can see now that the difference between wheat costing $1.50 a bushel and $4 a bushel in the overall price of a loaf of bread is negligible. It sure as hell doesn't make a difference. We get the same finger pointed at us when the people get a 25-cent-an-hour raise as the farmer does when his prices go up somewhat."

We can anticipate that as the class polarization deepens in this country, as the social and economic crisis sharpens, as the threatening catastrophe that we've been talking about looms ever closer, ruling class demagogy will be stepped up. To counter these lies, farmers and workers must fix the spotlight on the banks and the food trusts—exposing their secret dealings, exorbitant profit-gouging, and total disregard for human needs.

How this can be done is described by Trotsky in *The Transitional Program*:

"The peasant, artisan, or small merchant, unlike the industrial worker or office or civil service employee, cannot demand a wage increase corresponding to the increase in prices. The official struggle of the government with high prices is only a deception of the masses. But the farmers, artisans, and merchants, in their capacity of consumers, can step into the politics of price-fixing shoulder to shoulder with the workers. To the capitalist's lamentations about costs of production, transport, and trade, the consumers answer: 'Show us your books; we demand control over the fixing of prices.' The organs of this control should be *committees on prices*, made up of delegates from the factories, trade unions, cooperatives, farmers' organizations, the 'little man' of the city, housewives, etc. By this means the workers will be able to prove to the farmers that the real reason for high prices is not high wages but the exorbitant profits of the capitalists and the overhead expenses of capitalist anarchy."

These price committees working jointly with committees of bank employees would also demand to inspect the books of the banks, whose soaring interest rates devastate the farmer, the homeowner, and those who buy on installment plans. Dragging into the open the truth about how the banks rob the farmer from every side is necessary if farmers and workers are to take into their hands control of transport, credit, and distribution operations affecting agriculture.

Environment, health, and safety
Another important problem farmers are fighting against is the destruction of the environment and all that this implies for health and safety. It's quite natural that they do so. The men and women who work the land, who are dependent on the soil, water, and clean air, have a close knowledge of a dependence on the environment. As people who

work for a living all the questions of health and safety are of major importance.

One protest is the battle in Minnesota against the construction of high tension wires on farmlands. Electric companies built a big power plant in North Dakota and are running their wires to the Twin Cities. There is evidence that the high voltage may cause damage to the central nervous system, blood chemistry, and bone tissue of human beings and farm animals. They are such a hazard that the electric companies weren't permitted to run them through many wildlife areas because the deer wouldn't even walk under them.

The matter of democracy is also involved because the farmers were never consulted nor was it explained to them why these had to be built. This in spite of the fact that a couple of these electric companies are "co-ops" in which the farmers are formally members.

These protests which have been going on for several years involve hundreds of farmers. Farmers have torn down towers and organized big demonstrations. When the protests started the response of the government was to call up the biggest mobilization of state troopers in Minnesota history. Over a hundred farmers were arrested. The police have occupied farm fields to guard construction crews.

Most of these farmers have never before been involved in protests. But as a result of their activity they are becoming conscious of other struggles. For example, they recently sent representatives to an antinuclear power conference in South Dakota and sent a contingent to an antinuclear demonstration in the Twin Cities.

The nuclear issue is an important one for farmers as the near melt-down at Three Mile Island highlighted. The biggest industry in the Harrisburg area is the dairy industry and the release of Iodine 131 directly affected them. It settled over their fields and milk from their cows was totally unsaleable during that period. This disaster also served to bring to the surface a lot of information about the deadly effects of fall-out emitted from the normal functioning of the plant. During the height of the Three Mile Island disaster, an article appeared in the *Philadelphia Enquirer* which included interviews with farmers.

One woman explained what had happened in the months preceding this disaster. "We lost two 300-pound steers, one in December and one in January. Then we lost the cats—four of them died in one week in December. The electric company keeps assuring us there is no leakage."

Another farmer said, "Last year I lost twelve dogs. They died of cancer. That never used to happen. The wild game started to disappear around here, and I must have lost 70 to 80 head of cattle, and the vet wasn't able to tell me why they died."

Jane Lee who lives on a farm four miles west of the plant had a similar story. "A cow died of cancer," she began. "Then a calf died at seven days old, and another calf was born dead. Also, a goat aborted. We had four ducks and altogether they laid 70 eggs. Only seven hatched and one duck was born deformed." That's the effects of the normal, run-of-the-mill fall-out from these plants! The full report on what the effects of the greater release of radiation during the near melt-down is not yet in.

Someone might ask, "If farmers are so concerned with the environment why do they keep using dangerous pesticides and herbicides, and nitrate fertilizers that run off into the streams polluting them?" The answer is that it's not the working farmers that are responsible. It's those who exploit the farmers. It's those who keep raking off big profits by driving the farmers to farm intensively, to squeeze everything they can out of every acre. It's Dow, Dupont, Tenneco and other giant chemical corporations that manufacture herbicides and pesticides. The usual pattern is for a new pesticide to be put on the market by one of these companies and in four or five years, after selling large quantities, it will be pulled off the market. This often happens after some government investigation has proved that it is too dangerous. So another super-duper chemical will be put on the market and extensively promoted and the whole process will be repeated.

Those who exploit the farmers totally disregard the health of the farmers and the quality of our food. Farmers have a big stake in developing safe methods of dealing with pests and weeds, and new techniques for dealing with erosion runoff, and so on.

On this question we can learn a lot from the Cuban revolution. In a speech Fidel Castro gave in 1967 to a conference of private farmers, he said:

"Every time I travel through the mountains I suffer terribly. I suffer terribly because I've seen the colossal destruction that man has wrought in the mountains. Every time I travel through the Sierra Maestra, the Escambray mountains, the mountains of the Second Front, many of the regions of the country, I can't help feeling sad to see how man has been destroying nature.

"And that nature is the nature that other generations will have to live off in 20, 50, or a hundred years; it's the nature that double, triple, four, five, ten times more population will have to live off in the future. We ask ourselves whether this generation of Cubans has the right to destroy nature? Do they have the right to leave future generations barren rock? And naturally we have to admit that they don't have that right.

"But we also ask ourselves whether that Cuban, that farmer is to blame for having been forced to commit that crime against nature? No, no! What was it that forced that human being, that man to climb to the top of the hill to cut down the forest, to burn the timber, to plant anything one or two years, while the rains came and carried off the topsoil? What forced him? Did he go there for the fun of it? Did he go there aware of what he was doing? He was forced to go there by an inhuman social regime, an exploiting social regime, a selfish regime."

Working farmers as fellow toilers

Our approach, the approach we think the labor movement should have toward farmers, is first of all to distinguish between the exploited farmer and the exploiting farmer, between the producer using family labor and the capitalist producer employing wage labor.

We approach the working farmer as fellow workers, as men and women who work for a living and help to produce the goods and services that make up the vast wealth of this country.

The ruling class attempts to lump farmers together to give the impression that all farmers have common interests, as businessmen, who are different from and opposed to labor.

We, on the other hand, approach working farmers as victims of capitalism. They are exploited. Their sons go to war. Their daughters are denied equal rights. They too suffer from the shortages and breakdowns capitalism inflicts on all working people.

So a central element of labor's program for agriculture must be the promise to working farmers that a workers government won't expropriate them; that it won't take away their land, machinery and livestock; and that it won't turn them out and make their farms state property. Of course, the big capitalist farms that presently exist will be expropriated and placed under workers management. But the independent producer, the exploited farmer, will remain owner of his farm as long as he himself believes it possible or necessary. This is not because we cherish the moral virtues of family farming; nor because we think private ownership in agriculture is necessarily more efficient from a technical and economic standpoint.

Our approach is totally *political*. It is designed to win the confidence and support of exploited farmers, to show we are reliable protectors of their vital interests. Without this position, working farmers can be driven into supporting the bourgeoisie, and even fascism, as the class polarization deepens.

In a major upheaval, in a breakdown of society, the farmer, if turned against the working class, can sabotage production and distribution, preventing workers from getting food. Think what would happen if food stopped coming into New York City just for one day. Even though there are very few farmers, they have the power to make it impossible for working people to take or to hold power. So the correct program for them is a life and death matter, certainly not a peripheral question.

The Stalinist policy of forced collectivization cannot be part of it. This dreadful bureaucratic course has nothing to do with the revolutionary Marxist approach to the farm question. It is the very opposite. The position of Marxism has been clear on this from the time of Marx and Engels to the adoption of the Transitional Program by the Fourth International. We can point to how the Bolsheviks under Lenin applied this policy in the early days of the Russian Revolution and codified it in the program of the Third International at its Second Congress in 1920.

The most recent and effective example we can refer to is the Cuban revolution. There they carried out an exemplary policy, totally different from the Stalinist policy of forced collectivization imposed in Eastern Europe, China and by Stalin in the Soviet Union. In Cuba the big landlords were nationalized and state farms and cooperatives were established. But deeds were given to thousands of small farmers, who were renters or squatters and taxes were eliminated. Castro, describing this in a speech in 1962, said:

"Now the counterrevolutionaries say to the farmers, this is socialism, and they are going to socialize the land. But we say this clearly to the small farmer: 'Don't believe those tales; this is socialism and for this very reason we are not going to take your land. Why? Because you, the farmer, are an ally of the working class, because you, the small farmer, do not exploit anybody: you work with the help of your family and you produce. The working class is not going to take your land away from you; on the contrary, the working class grants you loans, sends you doctors, builds roads for you, educates your children, buys your products, pays you good prices, and strives to give you the supplies you need.' This is what the worker says to the farmer."

And that's what they've done in Cuba.

In another speech to small farmers five years later Castro reaffirmed this promise. "We wonder if there will be small farmers in forty years. And the answer is that if in forty years farmers still exist who want to be alone, isolated, working there with a yoke of oxen, with a very low productivity, who prefer to stay that way, we'll leave them there even if it's 40, 50, 100 years. Does that mean then that this will last forever? No. It won't last forever, but that won't be because of any law of any kind. It won't last forever because of the incredible, colossal development in the agriculture of this country, in Cuban society, because of the tremendous development of technology, because of the fantastic development of social and educational programs."

Like the revolutionary Cuban leadership, we American socialists promise not to expropriate the working farmer. This pledge isn't a formality or tongue-in-cheek concession to get them to join with us in the struggle for power. And then, after state power is taken by the working masses, to let them get along as best they can. Such a position must be backed up by a positive commitment, an energetic policy of aiding the farmers, offering low-interest credit, elimination of all taxes, free medical care, and full retirement pensions. The policy must be one of helping small farmers, even if, as Castro says, "it's 40, 50, 100 years."

Only through a common struggle for a workers government, a government truly representative of all working people, of all the oppressed and exploited, can working farmers be saved from utter ruin, degradation, and demoralization under monopoly capitalism. Their only future is in an alliance with labor's struggle for power. They have a genuine stake in having the unions break from capitalist politics and establish a labor party.

The history of local labor party formations in this country in the 1920s and 1930s shows that working farmers participated too. In Minnesota, because of the weight and initiating role of farmers, it was, in fact, called the Farmer-Labor Party. That development may not be repeated, at least in the same way, because of the changed relationship of forces, but working farmers will most certainly be a welcome part of the labor party that is going to be formed in this country and, as Trotsky urged, among its candidates will be working farmers. To forge a firm political alliance between labor and working farmers the labor party will have to put forward a program for farmers and counterpose it to those of the capitalist parties.

Finally, in the struggle against capitalism and in the transition to socialism, we do more than state that the interests of working farmers, of all the exploited, will be protected. We appeal to farmers of this country in still another way. We say that a workers government will urge farmers to become warriors in the battle against hunger throughout the world, to join the fight to end famines once and for all.

Working farmers are not self-interested profiteers, isolationists, people who can't read, write, or think about matters of world importance. They are debt slaves, men and women who know what exploitation is. And they will respond enthusiastically to the internationalist approach socialists project. Farmers like to produce, they want to produce, it's what they're in business to do, and it's what they know how to do. All farmers talk with pride of the immense productive capacity of American agriculture, of how many people they can feed. And the concept of overproduction leading to burning of crops and needless restrictions is an evil curse for them. It renders their efforts useless.

We will not wait for a workers government to talk about all this. We should talk about it and campaign around it now. We should make it part of our program and inspire farmers and fellow workers with it. Because the vast land resources and the unprecedented technology of the United States can make this country truly a granary to serve the peoples of the world rather than starving them.

Summary

Several comrades passed me notes asking for suggestions on what to read regarding the Marxist view of the farm question. Here are a few references I think are helpful:

1. *The Transitional Program for Socialist Revolution*, by Leon Trotsky (Pathfinder Press, 1977) [2009 printing]. Comrades are particularly referred to pp. 124–125; 163–166; and 251–256.

2. "The Peasant Question in France and Germany," by Frederick Engels. In the two-volume *Selected Works of Marx and Engels*, (Progress Publishers, 1962). This article is found in Vol. 2, p. 381. In the more recent three-volume edition it is found in Vol. 3, p. 457. Written in the last year of his life, this article deals with the question of farmers in two imperialist countries near the turn of the century. It is especially helpful in sorting out the different strata in the rural population and explaining why forced collectivization would be a disastrous political error.

3. *Alliance of the Working Class and the Peasantry* (Progress Publishers), by Lenin. This is a collection of articles and speeches by Lenin on the agrarian question. Those after mid-1917 are by far the most useful and give a very clear picture how a revolutionary Marxist leadership dealt with this question in the struggle for power and in consolidating a workers state. Perhaps the most important document drafted by Lenin is the "Draft Theses on the Agrarian Question" adopted by the Second Congress of the Comintern in 1920.

4. There are a couple of recently published books on U. S. agriculture that comrades will also find of interest. One is *Dollar Harvest*, by Samuel R. Berger. It is a $5.00 paperback published by the American Agriculture Movement and is an exposé of the Farm Bureau. The other is *Merchants of Grain* by Dan Morgan published by Viking Press. Unfortunately it is only available in hardback at $14.95. This is an account of the role of the five giant monopolies that dominate the merchandising of grain in the world.

It is unfortunate that Castro's speeches on agriculture are either out of print or not readily available in English. The manner—even the very language—in which he discusses the questions of farmers and the policy to be followed by a workers state is reminiscent of Lenin. As Lenin did in regard to Russia, Castro starts from the specific features of agriculture in Cuba—the stratification of the rural population, how different sectors of agriculture are organized, how to win the political confidence of working farmers, etc. His speeches clearly approach the problem from a revolutionary standpoint.

On Cuba's agricultural policy, Ed [Shaw] raised the question of how much relevance it has for us in the United States. For example, wasn't one of the big problems in Cuba land reform, i.e. distribution of the land to the peasants?

It's true that there are a lot of differences between agriculture in Cuba and the U. S., but the situations are a bit more alike than might at first appear—more similar than with Tsarist Russia, for example.

The fact is that the main thrust of Cuban policy was not concerned with redistribution of the land, although that was an important part of it. It also included establishing cooperatives and state farms on the big plantations that primarily employed wage workers.

Finally, it should be understood that this report is only the beginning of our discussion of the farm question. Its purpose is to begin the process of getting the party to think and learn more about this question. Hopefully this can lead to a written resolution at some point.

Somebody raised in the discussion whether or not we had been correct in 1967 to change our transitional slogan: "For a workers and farmers government" to "For a workers government." I think this is one of the questions we'll want to consider further in the course of our discussion on the farm question.

We should bear in mind that we did not change our slogan as the result of a thorough discussion about American farmers—their social weight and role in the economy. Rather, the revision grew out of a discussion of how we could include other weighty allies of the working class, especially oppressed national minorities, in our governmental slogan. When we didn't come up with a satisfactory alternative on this, we decided to drop farmers from our governmental demand in consideration of the decrease in the numerical size of the farm population.

Now we can rethink this question. But it's better to have our discussion and establish our position on American agriculture first. Then we'll be on firmer ground in the decision on our governmental slogan.

THE PEASANT QUESTION IN FRANCE AND GERMANY[1]
by Frederick Engels

The bourgeois and reactionary parties greatly wonder why everywhere among Socialists the peasant question has now suddenly been placed upon the order of the day. What they should be wondering at, by rights, is that this has not been done long ago. From Ireland to Sicily, from Andalusia to Russia and Bulgaria, the peasant is a very essential factor of the population, production and political power. Only two regions of Western Europe form an exception. In Great Britain proper big landed estates and large-scale agriculture have totally displaced the self-supporting peasant; in Prussia east of the Elbe the same process has been going on for centuries; here too the peasant is being increasingly "turned out" or at least economically and politically forced into the background.

The peasant has so far largely manifested himself as a factor of political power only by his apathy, which has its roots in the isolation of rustic life. This apathy on the part of the great mass of the population is the strongest pillar not only of the parliamentary corruption in Paris and Rome but also of Russian despotism. Yet it is by no means insuperable. Since the rise of the working-class movement in Western Europe, particularly in those parts where small peasant holdings predominate, it has not been particularly difficult for the bourgeoisie to render the socialist workers suspicious and odious in the minds of the peasants as *partageux*, as people who want to "divide up", as lazy greedy city dwellers who have an eye on the property of the peasants. The hazy socialistic aspirations of the revolution of February 1848 were rapidly disposed of by the reactionary ballots of the French peasantry; the peasant, who wanted peace of mind, dug up from his treasured memories the legend of Napoleon, the emperor of the peasants, and created the Second Empire. We all know what this one feat of the peasants cost the people of France; it is still suffering from its aftermath.

But much has changed since then. The development of the capitalist form of production has cut the life-strings of small production in agriculture; small production is irretrievably going to rack and ruin. Competitors in North and South America and in India have swamped the European market with their cheap grain, so cheap that no domestic producer can compete with it. The big landowners and small peasants alike see ruin staring them in the face. And since they are both owners of land and country folk, the big landowners assume the role of champions of the interests of the small peasants, and the small peasants by and large accept them as such.

Meanwhile a powerful socialist workers' party has sprung up and developed in the West. The obscure presentiments and feelings dating back to the February Revolution have become clarified and acquired the broader and deeper scope of a programme that meets all scientific requirements and contains definite tangible demands; and a steadily growing number of Socialist deputies fight for these demands in the German, French and Belgian parliaments. The conquest of political power by the Socialist Party has become a matter of the not too distant future. But in order to conquer political power this party must first go from the towns to the country, must become a power in the countryside. This party, which has an advantage over all

---

Published in Karl Marx and Frederick Engels, *Selected Works* (Moscow: Progress Publishers, 1970), vol. 3, pp. 457–78.

others in that it possesses a clear insight into the interconnections between economic causes and political effects and long ago descried the wolf in the sheep's clothing of the big landowner, that importunate friend of the peasant—may this party calmly leave the doomed peasant in the hands of his false protectors until he has been transformed from a passive into an active opponent of the industrial workers? This brings us right into the thick of the peasant question.

I

The rural population to which we can address ourselves consists of quite different parts, which vary greatly with the various regions.

In the west of Germany, as in France and Belgium, there prevails the small-scale cultivation of small-holding peasants, the majority of whom own and the minority of whom rent their parcels of land.

In the northwest—in Lower Saxony and Schleswig-Holstein—we have a preponderance of big and middle peasants who cannot do without male and female farm servants and even day labourers. The same is true of part of Bavaria.

In Prussia east of the Elbe and in Mecklenburg we have the region of big landed estates and large-scale cultivation with hinds, cotters and day labourers, and in between small and middle peasants in relatively unimportant and steadily decreasing proportion.

In central Germany all these forms of production and ownership are found mixed in various proportions, depending upon the locality, without the decided prevalence of any particular form over a large area.

Besides there are localities varying in extent where the arable land owned or rented is insufficient to provide for the subsistence of the family, but can serve only as the basis for operating a domestic industry and enabling the latter to pay the otherwise incomprehensibly low wages that ensure the steady sale of its products despite all foreign competition.

Which of these subdivisions of the rural population can be won over by the Social-Democratic Party? We, of course, investigate this question only in broad outline; we single out only clear-cut forms. We lack space to give consideration to intermediate stages and mixed rural populations.

Let us begin with the small peasant. Not only is he, of all peasants, the most important for Western Europe in general, but he is also the critical case that decides the entire question. Once we have clarified in our minds our attitude to the small peasant we have all the data needed to determine our stand relative to the other constituent parts of the rural population.

By small peasant we mean here the owner or tenant—particularly the former—of a patch of land no bigger, as a rule, than he and his family can till, and no smaller than can sustain the family. This small peasant, just like the small handicraftsman, is therefore a toiler who differs from the modern proletarian in that he still possesses his instruments of labour; hence a survival of a past mode of production. There is a threefold difference between him and his ancestor, the serf, bondman or, quite exceptionally, the free peasant liable to rent and feudal services. First, in that the French Revolution freed him from the feudal services and dues that he owed to the landlord and in the majority of cases, at least on the left bank of the Rhine, assigned his peasant farm to him as his own free property.

Secondly, in that he lost the protection of and the right to participate in the self-administering Mark community, and hence his share in the emoluments of the former common Mark. The common Mark was whisked away partly by the erstwhile feudal lord and partly by enlightened bureaucratic legislation patterned after Roman law. This deprives the small peasant of modern times of the possibility of feeding his draft animals without buying fodder. Economically, however, the loss of the emoluments derived from the Mark by far outweighs the benefits accruing from the abolition of feudal services. The number of peasants unable to keep draft animals of their own is steadily increasing.

Thirdly, the peasant of today has lost half of his former productive activity. Formerly he and his family produced, from raw material he had made himself, the greater part of the industrial products that he needed; the rest of what he required was supplied by village neighbours who plied a trade in addition to farming and were paid mostly in articles of exchange or in reciprocal services. The family, and still more the village, was self-sufficient, produced almost everything it needed. It was natural economy almost unalloyed; almost no money was necessary. Capitalist production put an end to this by its money economy and large-scale industry. But if the Mark emoluments represented one of the basic conditions of his existence, his industrial side line was another. And thus the peasant sinks ever lower. Taxes, crop failures, divisions of inheritance and litigations drive one peasant after another into the arms of the usurer; the indebtedness becomes more and more general and steadily increases in amount in each case—in brief, our small peasant, like every other survival of a past mode of production, is hopelessly doomed. He is a future proletarian.

As such he ought to lend a ready ear to socialist propaganda. But he is prevented from doing so for the time being by his deep-rooted sense of property. The more difficult it is for him to defend his endangered patch of land the more desperately he clings to it, the more he regards the Social-Democrats, who speak of transferring landed property to the whole of society, as just as dangerous a foe as the usurer and lawyer. How is Social-Democracy to overcome this prejudice? What can it offer to the doomed small peasant without becoming untrue to itself?

Here we find a practical point of support in the agrarian programme of the French Socialists of the Marxian trend, a programme which is the more noteworthy as it comes from the classical land of small-peasant economy.

The Marseilles Congress of 1892 adopted the first agrarian programme of the Party. It demands for propertyless rural *workers* (that is to say, day labourers and hinds): minimum wages fixed by trade unions and community councils; rural trade courts consisting half of workers; prohibition of the sale of common land; and the leasing of public domain land to communities which are to rent all this land, whether owned by them or rented, to associations of propertyless families of farm labourers for common cultivation, on condition that the employment of wage-workers be prohibited and that the communities exercise control; old-age and invalid pensions, to be defrayed by means of a special tax on big landed estates.

For the *small peasants,* with special consideration for tenant farmers, purchase of machinery by the community to be leased at cost price to the peasants; the formation of peasant co-operatives for the purchase of manure, drain-pipes, seed, etc., and for the sale of the produce; abolition of the real estate transfer tax if the value involved does not exceed 5,000 francs; arbitration commissions on the Irish pattern to reduce exorbitant rentals and compensate quitting tenant farmers and sharecroppers (*métayers*) for appreciation of the land due to them; repeal of Article 2102 of the Civil Code[2] which allows a landlord to distrain on the crop, and the abolition of the right of creditors to levy on growing crops; exemption from levy and distraint of a definite amount of farm implements and of the crop, seed, manure, draft animals, in short, whatever is indispensable to the peasant for carrying on his business; revision of the general cadastre, which has long been out of date, and until such time a local revision in each community; lastly, free instruction in farming, and agricultural experimental stations.

As we see, the demands made in the interests of the peasants—those made in the interests of the workers do not concern us here for the time being—are not very far-reaching. Part of them has already been realised elsewhere. The tenants' arbitration courts follow the Irish prototype by express mention. Peasant co-operatives already exist in the Rhine provinces. The revision of the cadastre has been a constant pious wish of all liberals and even bureaucrats throughout Western Europe. The other points, too, could be carried into effect without any substantial impairment of the existing capitalist order. So much simply in characterisation of the programme. No reproach is intended; quite the contrary.

The Party did such a good business with this programme among the peasants in the most diverse parts of France that—since appetite comes with eating—one felt constrained to suit it still more to their taste. It was felt, however, that this would be treading on dangerous ground. How was the peasant to be helped, not the peasant as a future proletarian but as a present propertied peasant, without violating the basic principles of the general socialist programme? In order to meet this objection the new practical proposals were prefaced by a theoretical preamble, which seeks to prove that it is in keeping with the principles of socialism to protect small-peasant property from destruction by the capitalist mode of production although one is perfectly aware that this destruction is inevitable. Let us now examine more closely this preamble as well as the demands themselves, which were adopted by the Nantes Congress in September of this year.

The preamble begins as follows:

> Whereas according to the terms of the general programme of the Party producers can be free only in so far as they are in possession of the means of production;
>
> Whereas in the sphere of industry these means of production have already reached such a degree of capitalist centralisation that they can be restored to the producers only in collective or social form, but in the sphere of agriculture—at least in present-day France—this is by no means the case, the means of production, namely, the land, being in very many localities still in the hands of the individual producers themselves as their individual possession;
>
> Whereas even if this state of affairs characterised by small-holding ownership is irretrievably doomed *(est fatalement appelé à disparaître)*, still it is not for socialism to hasten this doom, as its task does not consist in separating property from labour but, on the contrary, in uniting both of these factors of all production by placing them in the same hands, factors the separation of which entails the servitude and poverty of the workers reduced to proletarians;
>
> Whereas, on the one hand, it is the duty of socialism to put the agricultural proletarians again in possession—collective or social in form—of the great domains after expropriating their present idle owners, it is, on the other hand, no less its imperative duty to maintain the peasants themselves tilling their patches of land in possession of the same as against the fisk, the usurer and the encroachments of the newly-arisen big landowners;
>
> Whereas it is expedient to extend this protection also to the producers who as tenants or sharecroppers *(métayers)* cultivate the land owned by others and who, if they exploit day labourers, are to a certain extent compelled to do so because of the exploitation to which they themselves are subjected—
>
> Therefore the Workers' Party—which unlike the anarchists does not count on an increase and spread of poverty for the transformation of the social order but expects labour and society in general to be emancipated only by the organisation and concerted efforts of the workers of both country and town, by their taking possession of the government and legislation—has adopted the following agrarian programme in order thereby to bring together all the elements of rural production, all occupations which by virtue of various rights and titles utilise the national soil, to wage an identical struggle against the common foe: the feudality of landownership.

Now for a closer examination of these "whereases".

To begin with, the statement in the French programme that freedom of the producers presupposes the possession of the means of production must be supplemented by those immediately following: that the possession of the means of production is possible only in two forms: either as individual possession, which form never and nowhere existed for the producers in general, and is daily being made more impossible by industrial progress; or as common possession, a form the material and intellectual preconditions of which have been established by the development of capitalist society itself; that therefore taking *collective* possession of the means of production must be fought for by all means at the disposal of the proletariat.

The common possession of the means of production is thus set forth here as the sole principal goal to be striven for. Not only in industry, where the ground has already been prepared, but in general, hence also in agriculture. According to the programme individual possession never and nowhere obtained generally for all producers; for that very reason and because industrial progress removes it anyhow, socialism is not interested in maintaining but rather in removing it; because where it exists and in so far as it exists it makes common possession impossible. Once we cite the programme in support of our contention we must cite the entire programme, which considerably modifies the proposition quoted in Nantes; for it makes the general historical truth expressed in it dependent upon the conditions under which alone it can remain a truth today in Western Europe and North America.

Possession of the means of production by the individual producers nowadays no longer grants these producers real freedom. Handicraft has already been ruined in the cities; in metropolises like London it has already disappeared entirely, having been superseded by large-scale industry, the sweatshop system and miserable bunglers who thrive on bankruptcy. The self-supporting small peasant is neither in the safe possession of his tiny patch of land nor is he free. He as well as his house, his farmstead and his few fields belong to the usurer; his livelihood is more uncertain than that of the proletarian, who at least does have tranquil days now and then, which is never the case with the eternally tortured debt slave. Strike out Article 2102 of the Civil Code, provide by law that a definite amount of a peasant's farm implements, cattle, etc., shall be exempt from levy and distraint; yet you cannot ensure him against an emergency in which he is compelled to sell his cattle "voluntarily", in which he must sign himself away body and soul to the usurer and be glad to get a reprieve. Your attempt to protect the small peasant in his property does not protect his liberty but only the particular form of his servitude; it prolongs a situation in which he can neither live nor die. It is, therefore, entirely out of place here to cite the first paragraph of your programme as authority for your contention.

The preamble states that in present-day France the means of production, that is, the land, is in very many localities still in the hands of individual producers as their individual possession; that, however, it is not the task of socialism to separate property from labour, but, on the contrary, to unite these two factors of all production by placing them in the same hands. As has already been pointed out, the latter in this general form is by no means the task of socialism. Its task is rather only to transfer the means of production to the producers as their *common possession.* As soon as we lose sight of this the above statement becomes directly misleading in that it implies that it is the mission of socialism to convert the present sham property of the small peasant in his fields into real property, that is to say, to convert the small tenant into an owner and the indebted owner into a debtless owner. Undoubtedly socialism is interested to see that the false semblance of peasant property should disappear, but not in this manner.

At any rate we have now got so far that the preamble can straightforwardly declare it to be the duty of socialism, indeed, its imperative duty,

> to maintain the peasants themselves tilling their patches of land in possession of the same as against the fisk, the usurer and the encroachments of the newly-arisen big landowners.

The preamble thus imposes upon socialism the imperative duty to carry out something which it had declared to be impossible in the preceding paragraph. It charges it to "maintain" the small-holding ownership of the peasants although it itself states that this form of ownership is "irretrievably doomed". What are the fisk, the usurer and the newly-arisen big landowners if not the instruments by means of which capitalist production brings about this inevitable doom? What means "socialism" is to employ to protect the peasant against this trinity we shall see below.

But not only the small peasant is to be protected in his property. It is likewise

> expedient to extend this protection also to the producers who as tenants or sharecroppers (*métayers*) cultivate the land owned by others and who, if they exploit day labourers, are to a certain extent compelled to do so because of the exploitation to which they themselves are subjected.

Here we are entering upon ground that is passing strange. Socialism is particularly opposed to the exploitation of wage labour. And here it is declared to be the imperative duty of socialism to protect the French tenants when they *"exploit* day labourers", as the text literally states! And that because they are compelled to do so to a certain extent by "the exploitation to which they themselves are subjected"!

How easy and pleasant it is to keep on coasting once you are on the toboggan slide! When now the big and middle peasants of Germany come to ask the French Socialists to intercede with the German Party Executive to get the German Social-Democratic Party to protect them in the exploitation of their male and female farm servants, citing in support of their contention the "exploitation to which they themselves are subjected" by usurers, tax collectors, grain speculators and cattle dealers, what will they answer? What guarantee have they that our agrarian big landlords will not send them Count Kanitz (as he also submitted a proposal like theirs providing for a state monopoly of grain importation) and likewise ask for socialist protection of their exploitation of the rural workers, citing in support "the exploitation to which they themselves are subjected" by stock jobbers, money lenders and grain speculators?

Let us say here at the outset that the intentions of our French friends are not as bad as one would suppose. The above sentence, we are told, is intended to cover only a quite special case, namely, the following: In Northern France, just as in our sugar-beet districts, land is leased to the peasants subject to the obligation to cultivate beets, on conditions which are extremely onerous. They must deliver the beets to a stated factory at a price fixed by it, must buy definite seed, use a fixed quantity of prescribed fertiliser and on delivery are badly cheated into the bargain. We know all about this in Germany, as well. But if this sort of peasant is to be taken under one's wing this must be said openly and expressly. As the sentence reads now, in its unlimited general form, it is a direct violation not only of the French programme but also of the fundamental principle of socialism in general, and its authors will have no cause for complaint if this careless piece of editing is used against them in various quarters contrary to their intention.

Also capable of such misconstruction are the concluding words of the preamble according to which it is the task of the Socialist Workers' Party

> to bring together all the elements of rural production, all occupations which by virtue of various rights and titles utilise the national soil, to wage an identical struggle against the common foe: the feudality of landownership.

I flatly deny that the socialist workers' party of any country is charged with the task of taking into its fold, in addition to the rural proletarians and the small peasants, also the middle and big peasants and perhaps even the tenants of big estates, the capitalist cattle breeders and the other capitalist exploiters of the national soil. To all of them the feudality of landownership may appear to be a common foe. On certain questions we may make common cause with them and be able to fight side by side with them for definite aims. We can use in our Party individuals from every class of society, but have no use whatever for any groups representing capitalist, middle-bourgeois or middle-peasant interests. Here too what they mean is not as bad as it looks. The authors evidently never even gave all this a thought. But unfortunately they allowed themselves to be carried away by their zeal for generalisation and they must not be surprised if they are taken at their word.

After the preamble come the newly-adopted addenda to the programme itself. They betray the same cursory editing as the preamble.

The article providing that the communities must procure farming machinery and lease it at cost to the peasants is modified so as to provide that the communities are, in the first place, to receive state subsidies for this purpose and, secondly, that the machinery is to be placed at the disposal of the small peasants gratis. This further concession will not be of much avail to the small peasants, whose fields and mode of production permit of but little use of machinery.

Furthermore,

> substitution of a single progressive tax on all incomes upward of 3,000 francs for all existing direct and indirect taxes.

A similar demand has been included for many years in almost every Social-Democratic programme. But that this demand is raised in the special interests of the small peasants is something new and shows only how little its real scope has been calculated. Take Great Britain. There the state budget amounts to 90 million pounds sterling, of which 13½ to 14 million are accounted for by the income tax. The smaller part of the remaining 76 million is contributed by taxing business (post and telegraph charges, stamp tax), but by far the greater part of it by imposts on articles of mass consumption, by the constantly repeated clipping of small, imperceptible amounts totalling many millions from the incomes of all members of the population, but particularly of its poorer sections. In present-day society it is scarcely possible to defray state expenditures in any other way. Suppose the whole 90 million are saddled in Great Britain on the incomes of 120 pounds sterling=3,000 francs and in excess thereof by the imposition of a progressive direct tax. The average annual accumulation, the annual increase of the aggregate national wealth, amounted in 1865 to 1875, according to Giffen, to 240 million pounds sterling. Let us assume it now equals 300 million annually; a tax burden of 90 million would consume almost one-third of the aggregate accumulation. In other words, no government except a Socialist one can undertake any such thing. When the Socialists are at the helm there will be things for them to carry into execution alongside of which that tax reform will figure as a mere, and quite insignificant, settlement for the moment while altogether different prospects open up before the small peasants.

One seems to realise that the peasants will have to wait rather long for this tax reform so that "in the meantime" *(en attendant)* the following prospect is held out to them:

> Abolition of taxes on land for all peasants living by their own labour, and reduction of these taxes on all mortgaged plots.

The latter half of this demand can refer only to peasant farms *too big* to be operated by the family itself; hence it is again a provision in favour of peasants who "exploit day labourers".
Again:

> Hunting and fishing rights without restrictions other than such as may be necessary for the conservation of game and fish and the protection of growing crops.

This sounds very popular but the concluding part of the sentence wipes out the introductory part. How many rabbits, partridges, pikes and carps are there even today per peasant family in all the rural localities? Would you say more than would warrant giving each peasant just *one* day a year for free hunting and fishing?

> Lowering of the legal and conventional rate of interest—

hence renewed usury laws, a renewed attempt to introduce a police measure that has always failed everywhere for the last two thousand years. If a small peasant finds himself in a position where recourse to a usurer is the lesser evil to him, the usurer will always find ways and means of sucking him dry without falling foul of the usury laws. This measure could serve at most to soothe the small peasant but he will derive no advantage from it; on the contrary, it makes it more difficult for him to obtain credit precisely when he needs it most.

> Medical service free of charge and medicines at cost price—

this at any rate is not a measure for the special protection of the peasants. The German programme goes further and demands that medicine too should be free of charge.

> Compensation for families of reservists called up for military duty for the duration of their service—

this already exists, though most inadequately, in Germany and Austria and is likewise no special peasant demand.

> Lowering of the transport charges for fertiliser and farm machinery and products—

is on the whole in effect in Germany, and mainly in the interest—of the big landowners.

> Immediate preparatory work for the elaboration of a plan of public works for the amelioration of the soil and the development of agricultural production—

leaves everything in the realm of uncertitude and beautiful promises and is also above all in the interest of the big landed estates.

In brief, after the tremendous theoretical effort exhibited in the preamble the practical proposals of the new agrarian programme are even more unrevealing as to the way in which the French Workers' Party expects to be able to maintain the small peasants in possession of their small holdings, which, on its own testimony, are irretrievably doomed.

II

In one point our French comrades are absolutely right: No lasting revolutionary transformation is possible in France *against* the will of the small peasant. Only it seems to me they have not got the right leverage if they mean to bring the peasant under their influence.

They are bent, it seems, to win over the small peasant forthwith, possibly even for the next general elections. This they can hope to achieve only by making very risky general assurances in defence of which they are compelled to set forth even much more risky theoretical considerations. Then, upon closer examination, it appears that the general assurances are self-contradictory (promise to maintain a state of affairs which, as one declares oneself, is irretrievably doomed) and that the various measures are either wholly without effect (usury laws), or are general workers' demands or demands which also benefit the big landowners or finally are such as are of no great importance by any means in promoting the interests of the small peasants. In consequence, the directly practical part of the programme of itself corrects the erroneous initial part and reduces the apparently formidable grandiloquence of the preamble to actually innocent proportions.

Let us say it outright: in view of the prejudices arising out of their entire economic position, their upbringing and their isolated mode of life, prejudices nurtured by the bourgeois press and the big landowners, we can win the mass of the small peasants forthwith only if we make them a promise which we ourselves know we shall not be able to keep. That is, we must promise them not only to protect their property in any event against all economic forces sweeping upon them but also to relieve them of the burdens which already now oppress them: to transform the tenant into a free owner and to pay the debts of the owner succumbing to the weight of his mortgage. If we could do this we should again arrive at the point from which the present situation would necessarily develop anew. We shall not have emancipated the peasant but only given him a reprieve.

But it is not in our interests to win the peasant overnight only to lose him again on the morrow if we cannot keep our promise. We have no more use for the peasant as a Party member if he expects us to perpetuate his property in his small holding than for the small handicraftsman who would fain be perpetuated as a master. These people belong to the anti-Semites. Let them go to the anti-Semites and obtain from the latter the promise to salvage their small enterprises. Once they learn there what these glittering phrases really amount to and what melodies are fiddled down from the anti-Semitic heavens they will realise in ever-increasing measure that we who promise less and look for salvation in entirely different quarters are after all more reliable people. If the French had the strident anti-Semitic demagogy we have they would hardly have committed the Nantes mistake.

What, then, is our attitude towards the small peasantry? How shall we have to deal with it on the day of our accession to power?

To begin with, the French programme is absolutely correct in stating: that we foresee the inevitable doom of the small peasant but that it is not our mission to hasten it by any interference on our part.

Secondly, it is just as evident that when we are in possession of state power we shall not even think of forcibly expropriating the small peasants (regardless of whether with or without compensation), as we shall have to do in the case of the big landowners. Our task relative to the small peasant consists, in the first place, in effecting a transition of his private enterprise and private possession to co-operative ones, not forcibly but by dint of example and the proffer of social assistance for this purpose. And then of course we shall have ample means of showing to the small peasant prospective advantages that must be obvious to him even today.

Almost twenty years ago the Danish Socialists, who have only *one* real city in their country—Copenhagen—and therefore have to rely almost exclusively on peasant propaganda outside of it, were already drawing up such plans. The peasants of a village or parish—there are many big individual homesteads in Denmark—were to pool their land to form a single big farm in order to cultivate it for common account and distribute the yield in proportion to the land, money and labour contributed. In Denmark small landed property plays only a secondary role. But if we apply this idea to a region of small holdings we shall find that if these are pooled and the aggregate area cultivated on a large scale, part of the labour power employed hitherto is rendered superfluous. It is precisely this saving of labour that represents one of the main advantages of large-scale farming. Employment can be found for this labour power in two ways. Either additional land taken from big estates in the neighbourhood is placed at the disposal of the peasant co-operative or the peasants in question are provided with the means and the opportunity of engaging in industry as an accessory calling, primarily and as far as possible for their own use. In either case their economic position is improved and simultaneously the general social directing agency is assured the necessary influence to transform the peasant co-operative to a higher form, and to equalise the rights and duties of the co-operative as a whole as well as of its individual members with those of the other departments of the entire community. How this is to be carried out in practice in each particular case will depend upon the circumstances of the case and the conditions under which we take possession of political power. We may thus possibly be in a position to offer these cooperatives yet further advantages: assumption of their entire mortgage indebtedness by the national bank with a simultaneous sharp reduction of the interest rate; advances from public funds for the establishment of large-scale production (to be made not necessarily or primarily in money but in the form of required products: machinery, artificial fertiliser, etc.), and other advantages.

The main point is and will be to make the peasants understand that we can save, preserve their houses and fields for them only by transforming them into co-operative property operated co-operatively. It is precisely the individual farming conditioned by individual ownership that drives the peasants to their doom. If they insist on individual operation they will inevitably be driven from house and home and their antiquated mode of production superseded by capitalist large-scale production. That is how the matter stands. Now we come along and offer the peasants the opportunity of introducing large-scale production themselves, not for account of the capitalists but for their own, common account. Should it really be impossible to make the peasants understand that this is in their own interest, that it is the sole means of their salvation?

Neither now nor at any time in the future can we promise the small-holding peasants to preserve their individual property and individual enterprise against the overwhelming power of capitalist production. We can only promise them that we shall not interfere in their property relations by force, against their will. Moreover, we can advocate that the struggle of the capitalists and big landlords against the small peasants should be waged from now on with a minimum of unfair means and that direct robbery and cheating, which are practised only too often, be as far as possible prevented. In this we shall succeed only in exceptional cases. Under the developed capitalist mode of production nobody can tell where honesty ends and cheating begins. But always it will make a considerable difference whether public authority is on the side of the cheater or the cheated. We of course are decidedly on the side of the small peasant; we shall do everything at all permissible to make his lot more bearable, to facilitate his transition to the co-operative should he decide to do so, and even to make it possible for him to remain on his small holding for a protracted length of time to think the matter over, should he still be unable to bring himself to this decision. We do this not only because we consider the small peasant living by his own labour as virtually belonging to us, but also in the direct interest of the Party. The greater the number of peasants whom we can save from being actually hurled down into the proletariat, whom we can win to our side while they are still peasants, the more quickly and easily the social transformation will be accomplished. It will serve us nought to wait with this transformation until capitalist production has developed everywhere to its utmost consequences, until the last small handicraftsman and the last small peasant have fallen victim to capitalist large-scale production. The material sacrifice to be made for this purpose in the interest of the peasants and to be defrayed out of public funds can, from the point of view of capitalist economy, be viewed only as money thrown away, but it is nevertheless an excellent investment because it will effect a perhaps tenfold saving in the cost of the social reorganisation in general. In this sense we can, therefore, afford to deal very liberally with the peasants. This is not the place to go into details, to make concrete proposals to that end; here we can deal only with general principles.

Accordingly we can do no greater disservice to the Party as well as to the small peasants than to make promises that even only create the impression that we intend to preserve the small holdings permanently. It would mean directly to block the way of the peasants to their emancipation and to degrade the Party to the level of rowdy anti-Semitism. On the contrary, it is the duty of our Party to make clear to the peasants again and again that their position is absolutely hopeless as long as capitalism holds sway, that it is absolutely impossible to preserve their small holdings for them as such, that capitalist large-scale production is absolutely sure to run over their impotent antiquated system of small production as a train runs over a pushcart. If we do this we shall act in conformity with the inevitable trend of economic development, and this development will not fail to bring our words home to the small peasants.

Incidentally, I cannot leave this subject without expressing my conviction that the authors of the Nantes programme are also essentially of my opinion. Their insight is much too great for them not to know that areas now divided into small holdings are also bound to become common property. They themselves admit that small-holding ownership is destined to disappear. The report of the National Council drawn up by Lafargue and delivered at the Congress of Nantes likewise fully corroborates this view. It has been published in German in the Berlin *Sozialdemokrat* of October 18 of this year[3]. The contradictory nature of the expressions used in the Nantes programme itself betrays the fact that what the authors actually say is not what they want to say. If they are not understood and their statements misused, as actually has already happened, that is of course their own fault. At any rate, they will have to elucidate their programme and the next French congress revise it thoroughly.

We now come to the bigger peasants. Here as a result of the divisions of inheritance as well as of indebtedness and forced sales of land we find a variegated pattern of intermediate stages, from small-holding peasant to big peasant proprietor, who has retained his old patrimony intact or even added to it. Where the middle peasant lives among small-holding peasants his interests and views will not differ greatly from theirs; he knows from his own experience how many of his kind have already sunk to the level of small peasants. But where middle and big peasants predominate and the operation of the farms requires, generally, the help of male and female servants it is quite a different matter. Of course a workers' party has to fight, in the first place, on behalf of the wage-workers, that is, for the male and female servantry and the day labourers. It is unquestionably forbidden to make any promises to the peasants which include the continuance of the wage slavery of the workers. But as long as the big and middle peasants continue to exist as such they cannot manage without wage-workers. If it would, therefore, be downright folly on our part to hold out prospects to the small-holding peasants of continuing permanently to be such, it would border on treason were we to promise the same to the big and middle peasants.

We have here again the parallel case of the handicraftsmen in the cities. True, they are more ruined than the peasants but there still are some who employ journeymen in addition to apprentices or for whom apprentices do the work of journeymen. Let those of these master craftsmen who want to perpetuate their existence as such cast in their lot with the anti-Semites until they have convinced themselves that they get no help in that quarter either. The rest, who have realised that their mode of production is inevitably doomed, are coming over to us and, moreover, are ready in future to share the lot that is in store for all other workers. The same applies to the big and middle peasants. It goes without saying that we are more interested in their male and female servants and day labourers than in them themselves. If these peasants want to be guaranteed the continued existence of their enterprises we are in no position whatever to assure them of that. They must then take their place among the anti-Semites, peasant leaguers and similar parties who derive pleasure from promising everything and keeping nothing. We are economically certain that the big and middle peasants must likewise inevitably succumb to the competition of capitalist production and the cheap overseas corn, as is proved by the growing indebtedness and the everywhere evident decay of these peasants as well. We can do nothing against this decay except recommend here too the pooling of farms to form co-operative enterprises, in which the exploitation of wage labour will be eliminated more and more, and their gradual transformation into branches of the great national producers' co-operative with each branch enjoying equal rights and duties can be instituted. If these peasants realise the inevitability of the doom of their present mode of production and draw the necessary conclusions they will come to us and it will be incumbent upon us to facilitate to the best of our ability also their transition to the changed mode of production. Otherwise we shall have to abandon them to their fate and address ourselves to their wage-workers, among whom we shall not fail to find sympathy. Most likely we shall be able to abstain here as well from resorting to forcible expropriation, and as for the rest to count on future economic developments making also these harder pates amenable to reason.

Only the big landed estates present a perfectly simple case. Here we are dealing with undisguised capitalist production and no scruples of any sort need restrain us. Here we are confronted by rural proletarians in masses and our task is clear. As soon as our Party is in possession of political power it has simply to expropriate the big landed proprietors just like the manufacturers in industry. Whether this expropriation is to be compensated for or not will to a great extent depend not upon us but the circumstances under which we obtain power, and particularly upon the attitude adopted by these gentry, the big landowners, themselves. We by no means consider compensation as impermissible in any event; Marx told me (and how many times!) that in his opinion we would get off cheapest if we could buy out the whole lot of them. But this does not concern us here. The big estates thus restored to the community are to be turned over by us to the rural workers who are already cultivating them and are to be organised into co-operatives. They are to be assigned to them for their use and benefit under the control of the community. Nothing can as yet be stated as to the terms of their tenure. At any rate the transformation of the capitalist enterprise into a social enterprise is here fully prepared for and can be carried into execution overnight, precisely as in Mr. Krupp's or Mr. von Stumm's factory. And the example of these agricultural co-operatives would convince also the last of the still resistant small-holding peasants, and surely also many big peasants, of the advantages of co-operative, large-scale production.

Thus we can open up prospects here before the rural proletarians as splendid as those facing the industrial workers, and it can be only a question of time, and of only a very short time, before we win over to our side the rural workers of Prussia east of the Elbe. But once we have the East-Elbe rural workers a different wind will blow at once all over Germany. The actual semi-servitude of the East-Elbe rural workers is the main basis of the domination of Prussian Junkerdom and thus of Prussia's specific overlordship in Germany. It is the Junkers east of the Elbe who have created and preserved the specifically Prussian character of the bureaucracy as well as of the body of army officers—the Junkers, who are being reduced more and more to ruin by their indebtedness, impoverishment and parasitism at state and private cost and for that very reason cling the more desperately to the dominion which they exercise; the Junkers, whose haughtiness, bigotry and arrogance have brought the German Reich of the Prussian nation[4] within the country into such hatred—even when every allowance is made for the fact that at present this Reich is inevitable as the sole form in which national unity can now be attained—and abroad so little respect despite its brilliant victories. The power of these Junkers is grounded on the fact that within the compact territory of the seven old Prussian provinces—that is, approximately one-third of the entire territory of the Reich—they have at their disposal the landed property, which here brings with it both social and political power. And not only the landed property but, through their beet-sugar refineries and liquor distilleries, also the most important industries of this area. Neither the big landowners of the rest of Germany nor the big industrialists are in a similarly favourable position. Neither of them have a compact kingdom at their disposal. Both are scattered over a wide stretch of territory and compete among themselves and with other social elements surrounding them for economic and political predominance. But the economic foundation of this domination of the Prussian Junkers is steadily deteriorating. Here too indebtedness and impoverishment are spreading irresistibly despite all state assistance (and since Frederick II this item is included in every regular Junker budget). Only the actual semi-serfdom sanctioned by law and custom and the resulting possibility of the unlimited exploitation of the rural workers, still barely keep the drowning Junkers above water. Sow the seed of Social-Democracy among these workers, give them the courage and cohesion to insist upon their rights, and the glory of the Junkers will be at an end. The great reactionary power, which to Germany represents the same barbarous, predatory element as Russian tsardom does to the whole of Europe, will collapse like a pricked bubble. The "picked regiments" of the Prussian army will become Social-Democratic, which will result in a shift in power that is pregnant with an entire upheaval. But for this reason it is of vastly greater importance to win the rural proletariat east of the Elbe than the small peasants of Western Germany or yet the middle peasants of Southern Germany. It is here, in East-Elbe Prussia, that the decisive battle of our cause will have to be fought and for this very reason both government and Junkerdom will do their utmost to prevent our gaining access here. And should, as we are threatened, new violent measures be resorted to to impede the spread of our Party, their primary purpose will be to protect the East-Elbe rural proletariat from our propaganda. It's all the same to us. We shall win it nevertheless.

NOTES

1. Engels' *The Peasant Question in France and Germany* is a major Marxist work on the agrarian question. The immediate cause for writing this work was the attempt by Vollmar and other opportunists to make use of the discussion of the draft agrarian programme at the Frankfurt Congress of German Social-Democrats in 1894 in order to smuggle in an anti-Marxist theory on the socialist transformation of rich peasants, etc. Engels was also prompted to write this work by his striving to correct the mistakes committed by the French Socialists, who deviated from Marxism and made concessions to opportunism in their agrarian programme adopted in Marseilles in 1892 and supplemented in Nantes in 1894.

In addition Engels elucidates the revolutionary principles of the proletarian policy vis-à-vis the various groups of peasants and elaborates the idea of an alliance between the working class and the working peasantry.

2. By the Civil Code (the *Code Napoleon*) Engels implies the entire system of bourgeois law as represented by five codes (civil, civil procedure, commercial, criminal and criminal procedure) promulgated in the period 1804–10 under Napoleon Bonaparte. These codes were introduced in the western and south-western parts of Germany seized by Napoleonic France and continued to operate in the Rhine Province even after it was ceded to Prussia in 1815.

3. *Sozialdemokrat*—weekly of the Social-Democratic Party of Germany, which appeared in Berlin in 1894–95.

Paul Lafargue's report "Peasant Property and Economic Progress", mentioned by Engels, was published in the supplement to the newspaper on October 18, 1894.

4. Engels changes the name of the medieval Holy Roman Empire of the German Nation to emphasise that the unification of Germany was effected under Prussian supremacy and was attended by Prussification of the German lands.

PRELIMINARY DRAFT THESES ON THE AGRARIAN QUESTION
by V.I. Lenin

1. Only the urban and industrial proletariat, led by the Communist Party, can liberate the working masses of the countryside from the yoke of capital and landed proprietorship, from ruin and the imperialist wars that will inevitably break out again and again if the capitalist system remains. There is no salvation for the working masses of the countryside except in alliance with the Communist proletariat, and unless they give the latter devoted support in its revolutionary struggle to throw off the yoke of the landowners (the large landed proprietors) and the bourgeoisie.

On the other hand, the industrial workers cannot accomplish their epoch-making mission of emancipating mankind from the yoke of capital and from wars if they confine themselves to their narrow craft or trade interests and smugly restrict themselves to attaining an improvement in their own conditions, which may sometimes be tolerable in the petty-bourgeois sense. This is exactly what happens to the "labor aristocracy" of many advanced countries, who constitute the core of the so-called Socialist parties of the Second International. They are actually the bitter enemies and betrayers of socialism—petty-bourgeois chauvinists and agents of the bourgeoisie within the working-class movement.

The proletariat is a really revolutionary class and acts in a really socialist manner only when it comes out and acts as the vanguard of all the working and exploited people, as their leader in the struggle for the overthrow of the exploiters; this, however, cannot be achieved unless the class struggle is carried into the countryside, unless the rural working masses are united about the Communist Party of the urban proletariat, and unless they are trained by the proletariat.

2. The working and exploited people of the countryside, whom the urban proletariat must lead into the struggle or, at all events, win over, are represented in all capitalist countries by the following classes:

First, the agricultural proletariat, wage laborers (by the year, season, or day) who obtain their livelihood by working for hire at capitalist agricultural enterprises. The organization of this class (political, military, trade union, cooperative, cultural, educational, and so on) independently and separately from other groups of the rural population, the conduct of intensive propaganda and agitation among this class, and the winning of its support for the soviets and the dictatorship of the proletariat constitute the *fundamental* tasks of the Communist parties in all countries.

Second, the semiproletarians or peasants who till tiny plots of land, that is, those who obtain their livelihood partly as wage laborers at agricultural and industrial capitalist enterprises and partly by working their own or rented plots of land, which provide their families only with part of their means of subsistence. This group of the rural working population is very numerous in all capitalist countries. Its existence and special position are played down by the representatives of the bourgeoisie and by the Yellow "Socialists" belonging to the Second International, partly by deliber-

---

Reprinted from *Workers of the World and Oppressed Peoples, Unite! Proceedings of the Second Congress, 1920* (New York: Pathfinder, 1991), vol. 2, pp. 1212–26 [2013 printing]. There was substantial discussion and debate on the draft in the commission that prepared the report. In the resolution adopted by the congress, a number of changes were made in Lenin's draft, especially on the question of land distribution to the peasants. That resolution is also included in the *Second Congress* collection.

ately deceiving the workers and partly by blindly submitting to the routine of petty-bourgeois views and lumping together this group with the mass of the "peasantry."

This bourgeois method of duping the workers is to be seen mostly in Germany and in France, but also in America and other countries. If the work of the Communist Party is properly organized, this group will become its assured supporter, for the lot of these semiproletarians is a very hard one and they stand to gain enormously and immediately from soviet government and the dictatorship of the proletariat.

Third, the small peasantry, that is, the small-scale tillers who, either as owners or as tenants, hold small plots of land that enable them to satisfy the needs of their families and their farms and who do not hire outside labor. This stratum, as such, undoubtedly stands to gain by the victory of the proletariat, which will fully and immediately bring it (a) deliverance from the necessity of paying the large landowners rent or a share of the crop (for example the *métayers* in France, also in Italy and other countries); (b) deliverance from mortgages; (c) deliverance from the numerous forms of oppression by and dependence on the large landowners (forest lands and their use, etc.); (d) immediate aid for their farms from the proletarian state (the use of the agricultural implements and part of the buildings on the large capitalist farms confiscated by the proletariat and the immediate conversion, by the proletarian state, of the rural cooperative societies and agricultural associations from organizations that under capitalism served above all the rich and middle peasants into organizations that will primarily assist the poor, that is, proletarians, semiproletarians, small peasants, etc.), and many other things.

At the same time the Communist Party must clearly realize that during the transitional period from capitalism to communism, that is, during the dictatorship of the proletariat, this stratum, or at all events part of it, will inevitably vacillate toward unrestricted freedom of trade and the free enjoyment of the rights of private property. That is because this stratum, which, if only in a small way, is a seller of articles of consumption, has been corrupted by profiteering and by proprietary habits. However, if a firm proletarian policy is pursued, and if the victorious proletariat deals very resolutely with the large landowners and the large peasants, this stratum's vacillation cannot be considerable and cannot alter the fact that, on the whole, it will side with the proletarian revolution.

3. Taken together, the three groups enumerated above constitute the majority of the rural population in all capitalist countries. That is why the success of the proletarian revolution is fully assured, not only in the cities but in the countryside as well. The reverse view is widespread. However, it persists only, first, because of the deception systematically practiced by bourgeois science and statistics, which do everything to gloss over both the gulf that separates the above-mentioned classes in the countryside from the exploiters—the landowners and capitalists—and that which separates the semiproletarians and small peasants from the large peasants.

Second, it persists because of the inability and unwillingness of the heroes of the Yellow Second International and of the "labor aristocracy" in the advanced countries, which has been corrupted by imperialist privileges, to conduct genuinely proletarian revolutionary work of propaganda, agitation, and organization among the rural poor. The attention of the opportunists has always been and still is wholly concentrated on inventing theoretical and practical compromises with the bourgeoisie, including the large and middle peasants (who are dealt with below), and not on the revolutionary overthrow of the bourgeois government and the bourgeoisie by the proletariat.

It persists, third, because of the obstinate refusal to understand—so obstinate as to be equivalent to a prejudice (connected with all the other bourgeois-democratic and parliamentary prejudices)—a truth that has been fully proved by Marxist theory and fully corroborated by the experience of the proletarian revolution in Russia, namely, that although the three enumerated categories of the rural population—who are incredibly downtrodden, disunited, crushed, and doomed to semibarbarous conditions of existence in all countries, even the most advanced—are economically, socially, and culturally interested in the victory of socialism, they are capable of giving resolute support to the revolutionary proletariat only *after* the latter has won political power, only *after* it has resolutely dealt with the large landowners and capitalists, and only *after* these downtrodden people see in *practice* that they have an organized leader and champion, strong and firm enough to assist and lead them and to show them the right path.

4. In the economic sense, one should understand by "middle peasants" those small farmers who (a) either as owners or tenants, hold plots of land that are also small but, under capitalism, are sufficient not only to provide, as a general rule, a meager subsistence for the family and the bare minimum needed to maintain the farm, but also produce a certain surplus that may, in good years at least, be converted into capital; (b) quite frequently (for example, one farm out of two or three) resort to the employment of hired labor.

A concrete example of the middle peasants in an advanced capitalist country is provided by the group of farms of five to ten hectares [twelve to twenty-five acres] in Germany, in which, according to the census of 1907, the number of farms employing hired laborers is about one-third of the total number of farms in this group. In France, where the cultivation of special crops is more developed—for example, grape growing, which requires a very large amount of labor—this group probably employs outside hired labor to a somewhat greater extent.

The revolutionary proletariat cannot set itself the task—at least not in the immediate future or in the initial period of the dictatorship of the proletariat—of winning over this stratum, but must confine itself to the task of neutralizing it, that is, rendering it neutral in the struggle between the proletariat and the bourgeoisie. This stratum inevitably vacillates between these two forces. In the beginning of the new epoch and in the developed capitalist countries, it will, in the main, incline toward the bourgeoisie. That is because the world outlook and the sentiments of the property owners are prevalent among this stratum, which has a direct interest in profiteering, in "freedom" of trade and in property, and stands in direct antagonism to the wageworkers.

By abolishing rent and mortgages, the victorious proletariat will immediately improve the position of this stratum. In most capitalist countries, however, the proletarian state should not at once completely abolish private property. At all events, it guarantees both the small and the middle peasantry not only the preservation of their plots of land but also their enlargement to cover the total area they usually rented (the abolition of rent).

A combination of such measures with a ruthless struggle against the bourgeoisie fully guarantees the success of the policy of neutralization. The proletarian state must effect the transition to collective farming with extreme caution and only very gradually, by the force of example, without any coercion of the middle peasant.

5. The large peasants *(Grossbauern)* are capitalist entrepreneurs in agriculture, who as a rule employ several hired laborers and are connected with the "peasantry" only in their low cultural level, habits of life, and the manual labor they themselves perform on their farms. These constitute the biggest of the bourgeois strata, who are open and determined enemies of the revolutionary proletariat. In all their work in the countryside, the Communist parties must concentrate their attention mainly on the struggle against this stratum, on liberating the toiling and exploited majority of the rural population from the ideological and political influence of these exploiters, etc.

Following the victory of the proletariat in the cities, all sorts of manifestations of resistance and sabotage, as well as direct armed action of a counterrevolutionary character on the part of this stratum, are absolutely inevitable. The revolutionary proletariat must therefore immediately begin the ideological and organizational preparation of the forces necessary to completely disarm this stratum and, simultaneously with the overthrow of the capitalists in industry, to deal this stratum a most determined, ruthless, and smashing blow at the very first signs of resistance. For this purpose, the rural proletariat must be armed and village soviets organized, in which the exploiters must have no place, and in which proletarians and semiproletarians must be ensured predominance.

However, the expropriation even of the large peasants can in no way be made an immediate task of the victorious proletariat, because the material and especially the technical conditions, as well as the social conditions, for the socialization of such farms are still lacking. In individual and probably exceptional cases, those parts of their land that they rent out in small plots or that are particularly needed by the surrounding small-peasant population will be confiscated. The small peasants should also be guaranteed, on certain terms, the free use of part of the agricultural machinery belonging to the large peasants, etc. As a general rule, however, the proletarian state must allow the large peasants to retain their land, confiscating it only if they resist the power of the working and exploited people.

The experience of the Russian proletarian revolution, in which the struggle against the large peasantry was complicated and protracted by a number of special conditions, showed nevertheless that, when taught a severe lesson for the slightest attempt at resistance, this stratum is capable of loyally fulfilling the tasks set by the proletarian state, and even begins to be imbued, although very slowly, with respect for the government that protects all who work and is ruthless toward the idle rich.

The special conditions that, in Russia, complicated and retarded the struggle of the proletariat against the large peasants after it had defeated the bourgeoisie were, in the main, the following: after November 7, 1917 [October 25, old style], the Russian revolution passed through the stage of the "general democratic"—that is, basically the bourgeois-democratic—struggle of the peasantry as a whole against the landowners; the cultural and numerical weakness of the urban proletariat; and, lastly, the enormous distances and extremely poor means of communication.

Inasmuch as these retarding conditions do not exist in the advanced countries, the revolutionary proletariat of Europe and America should prepare far more energetically and achieve far more rapidly, resolutely, and successfully complete victory over the resistance of the large peasantry, completely depriving it of the slightest possibility of offering resistance. This is imperative because, until such a complete and absolute victory is achieved, the masses of the rural proletarians, semiproletarians, and small peasants cannot be brought to accept the proletarian state as a fully stable one.

6. The revolutionary proletariat must immediately and unreservedly confiscate all landed estates, those of the large landowners, who, in capitalist countries—directly or through their tenant farmers—systematically exploit wage labor and the neighboring small (and, not infrequently, part of the middle) peasantry, do not themselves engage in manual labor, and are in the main descended from the feudal lords (the nobles in Russia, Germany, and Hungary, the restored seigneurs in France, the lords in Britain, and the former slave owners in America), or are rich financial magnates, or else a mixture of both these categories of exploiters and parasites.

Under no circumstances is it permissible for Communist parties to advocate or practice compensating the large landowners for the confiscated lands, for under present-day conditions in Europe and America this would be tantamount to a betrayal of socialism and the imposition of new tribute upon the masses of working and exploited people, to whom the war has meant the greatest hardships, while it has increased the number of millionaires and enriched them.

As to the mode of cultivation of the land that the victorious proletariat confiscates from the large landowners, the distribution of that land among the peasantry for their use has been predominant in Russia, owing to her economic backwardness; it is only in relatively rare and exceptional cases that state farms have been organized on the former estates that the proletarian state runs at its own expense, converting the former wage laborers into workers for the state and members of the soviets, which administer the state. The Communist International is of the opinion that in the case of the advanced capitalist countries it would be correct to keep *most* of the large agricultural enterprises intact and to conduct them on the lines of the "state farms" in Russia.

It would, however, be grossly erroneous to exaggerate or to stereotype this rule and never to permit the free grant of *part* of the land that belonged to the expropriated expropriators to the neighboring small and sometimes middle peasants.

First, the objection usually raised to this, namely, that large-scale farming is technically superior, often amounts to an indisputable theoretical truth being replaced by the worst kind of opportunism and betrayal of the revolution. To achieve the success of this revolution, the proletariat should not shrink from a temporary decline in production, any more than the bourgeois opponents of slavery in North America shrank from a temporary decline in cotton production as a consequence of the Civil War of 1863–65.

What is most important to the bourgeois is production for the sake of production; what is most important to the working and exploited population is the overthrow of the exploiters and the creation of conditions that will permit the working people to work for themselves and not for the capitalists. It is the primary and fundamental task of the proletariat to ensure the proletarian victory and its stability. There can, however, be no stable proletarian government unless the middle peasantry is neutralized and the support is secured of a very considerable section of the small peasantry, if not all of them.

Second, not merely an increase but even the preservation of large-scale production in agriculture presupposes the existence of a fully developed and consciously revolutionary rural proletariat with considerable experience of trade union and political organization behind it. Where this condition does not yet exist, or where this work cannot expediently be entrusted to class-conscious and competent industrial workers, hasty attempts to set up large state-conducted farms can only discredit the proletarian government. Under such conditions, the utmost caution must be exercised and the most thorough preparations made when state farms are set up.

Third, in all capitalist countries, even the most advanced, there still exist survivals of medieval, semifeudal exploitation of the neighboring small peasants by the large landowners, as in the case of the *Instleute* [tenant farmers] in Germany, the *métayers* in France, and the sharecroppers in the United States (not only Negroes, who, in the southern states, are mostly exploited in this way, but sometimes whites too). In such cases it is incumbent on the proletarian state to grant the small peasants free use of the lands they formerly rented, since no other economic or technical basis exists and it cannot be created at one stroke.

The implements and stock of the large farms must be confiscated without fail and converted into state property, with the absolute condition that *after* the requirements of the large state farms have been met, the neighboring small peasants may have the use of these implements gratis, in compliance with conditions drawn up by the proletarian state.

In the period immediately following the proletarian revolution, it is absolutely necessary not only to confiscate the estates of the large landowners at once but also to deport or to intern them all as leaders of counterrevolution and ruthless oppressors of the entire rural population. However, with the consolidation of the proletarian power in the countryside as well as in the cities, systematic efforts should be made to employ (under the special control of highly reliable Communist workers) those forces within this class that possess valuable experience, know-how, and organizing skill, to build large-scale socialist agriculture.

7. The victory of socialism over capitalism and the consolidation of socialism may be regarded as ensured only when the proletarian state power, having completely suppressed all resistance by the exploiters and assured itself complete subordination and stability, has reorganized the whole of industry on the lines of large-scale collective production and on a modern technical basis (founded on the electrification of the entire economy). This alone will enable the cities to render such radical assistance, technical and social, to the backward and scattered rural population as will create the material basis necessary to boost the productivity of agricultural and of farm labor in general, thereby encouraging the small farmers by the force of example and in their own interests to adopt large-scale, collective, and mechanized agriculture.

Although nominally recognized by all socialists, this indisputable theoretical truth is in fact distorted by the opportunism prevalent in the Yellow Second International and among the leaders of the German and the British "Independents," the French Longuetists, etc. This distortion consists in attention being directed toward the relatively remote, beautiful, and rosy future; attention is deflected from the immediate tasks of the difficult practical transition and approach to that future. In practice, it consists in preaching a compromise with the bourgeoisie and a "class truce," that is, complete betrayal of the proletariat, which is now waging a struggle amid the unprecedented ruin and impoverishment created everywhere by the war, and amid the unprecedented enrichment and arrogance of a handful of millionaires resulting from that war.

It is in the countryside that a genuine possibility of a successful struggle for socialism demands, first, that all Communist parties should inculcate in the industrial proletariat a realization of the need to make sacrifices, and be prepared to make sacrifices so as to overthrow the bourgeoisie and consolidate proletarian power—since the dictatorship of the proletariat implies both the ability of the proletariat to organize and lead all the working and exploited people, and the vanguard's ability to make the utmost sacrifices and to display the utmost heroism to that end.

Second, success demands that, as a result of the workers' victory, the laboring and most exploited masses in the countryside achieve an immediate and considerable improvement in their conditions at the expense of the exploiters—for without that the industrial proletariat cannot get the support of the rural areas and, in particular, will be unable to ensure the supply of food for the cities.

8. The enormous difficulty of organizing and training for the revolutionary struggle the masses of rural working people, whom capitalism has reduced to a state of great wretchedness, disunity, and frequently semimedieval dependence, makes it necessary for the Communist parties to devote special attention to the strike struggle in the rural districts, give greater support to mass strikes by the agricultural proletarians and semiproletarians, and help develop the strike movement in every way.

The experience of the Russian revolutions of 1905 and of 1917, now confirmed and extended by the experience of Germany and other advanced countries, shows that the growing mass strike struggle (into which, under certain conditions, the small peasants can and should also be drawn) alone is capable of rousing the countryside from its lethargy, awakening the class consciousness of the exploited masses in the countryside, making them realize the need for class organization, and revealing to them in a vivid and practical manner the importance of their alliance with the urban workers.

This congress of the Communist International brands as traitors and renegades those Socialists—to be found, unfortunately, not only in the Yellow Second International but also in the three very important European parties that have withdrawn from that International—who are capable not only of remaining indifferent to the strike struggle in the countryside, but even (like Karl Kautsky) of opposing it on the grounds that it threatens to reduce the output of articles of consumption.

Neither programs nor the most solemn declarations are of any value whatever unless it is proved in practice, in deed, that the Communists and workers' leaders are able to place above everything else in the world the development and the victory of the proletarian revolution, and to make the greatest sacrifices for it, for otherwise there is no way out, no salvation from starvation, ruin, and new imperialist wars.

In particular, it should be pointed out that the leaders of the old Socialist movement and representatives of the "labor aristocracy"—who now often make verbal concessions to communism and even nominally side with it in order to preserve their prestige among the worker masses, who are rapidly becoming revolutionary—should be tested for their loyalty to the cause of the proletariat and their suitability for responsible positions in those spheres of work where the development of revolutionary consciousness and the revolutionary struggle is most marked, the resistance of the landowners and the bourgeoisie (the large peasants, the kulaks) most fierce, and the difference between the Socialist compromiser and the Communist revolutionary most striking.

9. The Communist parties must exert every effort to begin, as speedily as possible, to set up soviets of deputies in the countryside, and in the first place soviets of hired laborers and semiproletarians. Only if they are linked up with the mass strike struggle and with the most oppressed class can the soviets perform their functions and become consolidated enough to influence (and later to incorporate) the small peasants.

If, however, the strike struggle has not yet developed, and the agricultural proletariat is as yet incapable of strong organization owing both to the severe oppression by the landowners and the large peasants and to lack of support from the industrial workers and their unions, then the formation of soviets of deputies in the rural areas will require lengthy preparation by means of the organization of Communist cells, even if only small ones, intensified agitation—in which the demands of communism are enunciated in the simplest manner and illustrated by the most glaring examples of exploitation and oppression—and the arrangement of systematic visits of industrial workers to the rural districts, and so on.

REPORT ON WORK IN THE COUNTRYSIDE
by V.I. Lenin
March 23, 1919

[*Prolonged applause*] Comrades, I must apologise for having been unable to attend all the meetings of the committee elected by the Congress to consider the question of work in the countryside. My report will therefore be supplemented by the speeches of comrades who took part in the work of the committee from the very beginning. The committee finally drew up theses which were turned over to a commission and which will be reported on to you. I should like to dwell on the general significance of the question as it confronts us following the work of the committee and as, in my opinion, it now confronts the whole Party.

Comrades, it is quite natural that as the proletarian revolution develops we have to put in the forefront now one now another of the most complex and important problems of social life. It is perfectly natural that in a revolution which affects, and is bound to affect, the deepest foundations of life and the broadest mass of the population, not a single party, not a single government, no matter how close it may be to the people, can possibly embrace all aspects of life *at once*. And if we now have to deal with the question of work in the countryside, and in connection with this question to give prominence to the position of the middle peasants, there is nothing strange or abnormal in this from the standpoint of the development of the proletarian revolution in general. It is natural that the proletarian revolution had to begin with the fundamental relation between two hostile classes, the proletariat and the bourgeoisie. The principal task was to transfer power to the working class, to secure its dictatorship, to overthrow the bourgeoisie and to deprive them of the economic sources of their power, which are undoubtedly a hindrance to all socialist construction in general. Acquainted as we are with Marxism, not one of us has ever for a moment doubted the truth that, owing to the very economic structure of capitalist society, the deciding factor in that society can be either the proletariat or the bourgeoisie. We now see many former Marxists—from the Menshevik camp, for example—who assert that in a period of decisive struggle between the proletariat and the bourgeoisie *democracy in general* can prevail. This is what is said by the Mensheviks, who have come to a complete agreement with the Socialist-Revolutionaries. As though it is not the bourgeoisie themselves who create or abolish democracy as they find most convenient for themselves! And since that is so, there can be no question of democracy in general at a time of acute struggle between the bourgeoisie and the proletariat. It is astonishing how rapidly these Marxists, or pseudo-Marxists—our Mensheviks, for example—expose themselves, and how rapidly their true nature, the nature of petty-bourgeois democrats, comes to the surface.

Marx all his life fought most of all the illusions of petty-bourgeois democracy and bourgeois democracy. Marx scoffed most of all at empty talk of freedom and equality, when it serves as a screen for the freedom of the workers to starve to death, or the equality of one who sells his labour-power with the bourgeois who allegedly freely purchases the labour of the former in the open market as if from an equal, and so forth. Marx explains this in all his economic works. It may be said that the whole of Marx's *Capital* is devoted to explaining the truth that *the basic forces of capital-*

Published in V.I. Lenin, *Collected Works* (Moscow: Progress Publishers, 1965), vol. 29, pp. 198–215.

ist society are, and can only be, the bourgeoisie and the proletariat—bourgeoisie, as the builder of this capitalist society, as its leader, as its motive force, and the proletariat, as its gravedigger and as the only force capable of replacing it. You can hardly find a single chapter in any of Marx's works that is not devoted to this. You might say that all over the world the socialists of the Second International have vowed and sworn to the workers time and again that they understand this truth. But when matters reached the stage of the real and moreover decisive struggle for power between the proletariat and the bourgeoisie, we find that our Mensheviks and Socialist-Revolutionaries, as well as the leaders of the old socialist parties all over the world, forgot this truth and began to repeat in purely parrot fashion the philistine phrases about democracy in general.

Attempts are sometimes made to lend these words what is considered to be greater force by speaking of the "dictatorship of democracy". That is sheer nonsense. We know perfectly well from history that the dictatorship of the democratic bourgeoisie meant nothing but the suppression of the insurgent workers. That has been the case ever since 1848—at any rate, not later, and isolated examples may be found even earlier. History shows that it is precisely in a bourgeois democracy that a most acute struggle between the proletariat and the bourgeoisie proceeds widely and freely. We have had occasion to convince ourselves of this truth in practice. And the measures taken by the Soviet Government since October 1917 were distinguished by their firmness on all fundamental questions precisely because we have never departed from this truth and have never forgotten it. The issue of the struggle for supremacy waged against the bourgeoisie can be settled only by the dictatorship of one class—the proletariat. Only the dictatorship of the proletariat can defeat the bourgeoisie. Only the proletariat can overthrow the bourgeoisie. And only the proletariat can secure the following of the people in the struggle against the bourgeoisie.

However, it by no means follows from this—it would be a profound mistake to think it does—that in further building communism, when the bourgeoisie have been overthrown and political power is already in the hands of the proletariat, we can continue to carry on without the participation of the middle, intermediary elements.

It is natural that at the beginning of the revolution—the proletarian revolution—the whole attention of its active participants should be concentrated on the main and fundamental issue, the supremacy of the proletariat and the securing of that supremacy by a victory over the bourgeoisie—making it certain that the bourgeoisie cannot regain power. We are well aware that the bourgeoisie still enjoy the advantages derived from the wealth they possess in other countries or even the monetary wealth they sometimes possess in our own country. We are well aware that there are social elements who are more experienced than proletarians and who aid the bourgeoisie. We are well aware that the bourgeoisie have not abandoned the idea of returning to power and have not ceased attempting to restore their supremacy.

But that is by no means all. The bourgeoisie, who put forward most insistently the principle "my country is wherever it is good for me", and who, as far as money is concerned, have always been international—*the bourgeoisie internationally are at present still stronger than we are.* Their supremacy is being rapidly undermined, they are being confronted with such facts as the Hungarian revolution—about which we were happy to inform you yesterday and are today receiving confirming reports—and they are beginning to understand that their supremacy is shaky. They no longer enjoy freedom of action. But now, if you take into account the material means on the world scale, we cannot help admitting that in the material respect the bourgeoisie are at present still stronger than we are.

That is why nine-tenths of our attention and our practical activities were devoted, and had to be devoted, to this fundamental question—the overthrow of the bourgeoisie, the establishment of the power of the proletariat and the elimination of every possibility of the return of the bourgeoisie to power. That is absolutely natural, legitimate, and unavoidable, and very much in this respect has been successfully accomplished.

Now, however, we must decide the question of other sections of the population. We must—and this was our unanimous conclusion in the agrarian committee, and on this, we are convinced, all Party workers will agree, because we merely summed up the results of their observations—we must now decide *the question of the middle peasants* in its full magnitude.

Of course, there are people who, instead of reflecting on the course of our revolution, instead of pondering over the tasks now confronting us, instead of all this, make every step of the Soviet government a butt of derision and criticism of the type we hear from those gentlemen, the Mensheviks and the Right Socialist-Revolutionaries. These people have still not understood that they must make a choice between us and the bourgeois dictatorship. We have displayed great patience, even indulgence, towards these people. We shall allow them to enjoy our indulgence once more. But in the very near future we shall set a limit to our patience and indulgence, and if they do not make their choice, we shall tell them in all seriousness to go to Kolchak. [*Applause*] We do not expect particularly brilliant intellectual ability from such people. [*Laughter*] But it might have been expected that after experiencing the bestialities of Kolchak they ought to understand that we are entitled to demand that they should choose between us and Kolchak. If during the first few months that followed the October Revolution there were many naïve people who were stupid enough to believe that the dictatorship of the proletariat was something transitory and fortuitous, today even the Mensheviks and the Socialist-Revolutionaries ought to understand that there is something logically necessary in the struggle that is being waged before the onslaught of the whole international bourgeoisie.

Only two forces, in fact, have arisen: the dictatorship of the bourgeoisie and the dictatorship of the proletariat. Whoever has not learnt this from Marx, whoever has not learnt this from the works of all the great socialists, has never been a socialist, understood nothing about socialism, and has only called himself a socialist. We are allowing these people a short space for reflection and demand that they make their decision. I have mentioned them because they are now saying or will say: "The Bolsheviks have raised the question of the middle peasants; they want to make advances to them." I am very well aware that considerable space is given in the Menshevik press to arguments of this kind, and even far worse. We ignore such arguments, we never attach importance to the jabber of our adversaries. People who are still capable of running to and fro between the bourgeoisie and the proletariat may say what they please. We are following our own road.

Our road is determined above all by considerations of class forces. A struggle is developing in capitalist society between the bourgeoisie and the proletariat. As long as that struggle has not ended we shall give our keenest attention to fighting it out to the end. It has not yet been brought to the end. In that struggle much has already been accomplished. The hands of the international bourgeoisie are now no longer free. The best proof of this is that the Hungarian proletarian revolution has taken place. It is therefore clear that our constructive work in the countryside has already gone beyond the limits to which it was confined when everything was subordinated to the fundamental demand of the struggle for power.

This constructive work passed through two main phases. In October 1917 we seized power *together with the peasants as a whole*. This was a bourgeois revolution, inasmuch as the class struggle in the rural districts had not yet developed. As I have said, the real proletarian revolution in the rural districts began only in the summer of 1918. Had we not succeeded in stirring up this revolution our work would have been incomplete. The first stage was the seizure of power in the cities and the establishment of the Soviet form of government. The second stage was one which is fundamental for all socialists and without which socialists are not socialists, namely, to single out the proletarian and the semiproletarian elements in the rural districts and to weld them with the urban proletariat in order to wage the struggle against the bourgeoisie in the countryside. This stage is also in the main completed. The organisations we originally created for this purpose, the Poor Peasants' Committees, had become so consolidated that we found it possible to replace them by properly elected Soviets, i.e., to reorganise the village Soviets so as to make them the organs of class rule, the organs of proletarian power in the rural districts. Such measures as the law on socialist agrarian measures and measures for transition to socialist agriculture, which was passed not very long ago by the Central Executive

Committee and with which everybody, of course, is familiar, sum up our experience from the point of view of our proletarian revolution.

The main thing, the prime and basic task of the proletarian revolution, we have already accomplished. And precisely because we have accomplished it, a more complicated problem has come to the fore—*our attitude towards the middle peasants.* And whoever thinks that the fact that this problem is being brought to the fore is in any way symptomatic of a weakening of the character of our government, of a weakening of the dictatorship of the proletariat, that it is symptomatic of a change, however partial, however minute, in our basic policy, completely fails to understand the aims of the proletariat and the aims of the communist revolution. I am convinced that there are no such people in our Party. I only wanted to warn the comrades against people not belonging to the workers' party who will talk in this way, not because it follows from any system of ideas, but merely to spoil things for us and to help the whiteguards—or, to put it more simply, to incite against us the middle peasant, who is always vacillating, who cannot help vacillating, and who will continue to vacillate for a fairly long time to come. In order to incite the middle peasant against us they will say: "See, they are making advances to you! That means they have taken your revolts into account, they are beginning to wobble", and so on and so forth. All our comrades must be armed against agitation of this kind. And I am certain that they will be armed—provided we succeed now in having this question treated from the standpoint of the class struggle.

It is perfectly obvious that this fundamental problem—*how precisely to define the attitude of the proletariat towards the middle peasants*—is a more complex but no less urgent problem. Comrades, from the theoretical point of view, which has been mastered by the vast majority of the workers, this question presents no difficulty to Marxists. I will remind you, for instance, that in his book on the agrarian question, written at a time when he was still correctly expounding the teachings of Marx and was regarded as an indisputed authority in this field, Kautsky states in connection with the transition from capitalism to socialism that the task of a socialist party *is to neutralise the peasants,* i.e., to see to it that in the struggle between the proletariat and the bourgeoisie the peasant should remain neutral and should not be able to give active assistance to the bourgeoisie against us.

Throughout the extremely long period of the rule of the bourgeoisie, the peasants supported the power of the latter; they sided with the bourgeoisie. This will be understood if you consider the economic strength of the bourgeoisie and the political instruments of their rule. We cannot count on the middle peasant coming over to our side immediately. But if we pursue a correct policy, after a time these vacillations will cease and the peasant will be able to come over to our side.

It was Engels—who together with Marx laid the foundations of scientific Marxism, that is, the teachings by which our Party has always guided itself, and particularly in time of revolution—who already established the division of the peasants into small peasants, middle peasants, and big peasants, and this division holds good for the vast majority of European countries even today. Engels said: "Perhaps it will not be necessary to suppress even the big peasants by force everywhere." And that we might at any time use violence in relation to the middle peasants (the small peasant is our friend), that thought never occurred to any sensible socialist. That is what Engels said in 1894, a year before his death, when the agrarian question came to the fore. This point of view expresses a truth which is sometimes forgotten, but with which we are all in theory agreed. In relation to the landowners and the capitalists our aim is complete expropriation. *But we shall not tolerate any violence towards the middle peasants.* Even in regard to the rich peasants we do not say as resolutely as we say with regard to the bourgeoisie: absolute expropriation of the rich peasants and the kulaks. This distinction is observed in our programme. We say: the resistance and the counter-revolutionary efforts of the rich peasant must be suppressed. That is not complete expropriation.

The basic distinction that determines our attitude towards the bourgeoisie and the middle peasant—complete expropriation of the bourgeoisie and an alliance with the middle peasant who does not exploit others—this basic line is accepted by everybody in theory. But this line is not consistently followed in practice; they have not yet learnt to follow it in the localities. When, after having overthrown the bourgeoisie and consolidated its own power, the proletariat started from various angles to create a new society, the question of the middle peasant came to the fore. Not a single socialist in the world denied that the building of communism would take different courses in countries where large-scale agriculture prevails and in countries where small-scale agriculture prevails. That is an elementary truth, an ABC. And from this truth it follows that as we approach the problems of communist construction our principal attention must to a certain extent be concentrated precisely on the middle peasant.

Much will depend on how we define our attitude towards the middle peasant. Theoretically, that question has been solved; but we know perfectly well from our own experience that there is a difference between solving a problem theoretically and putting the solution into practice. We are now directly confronted with that difference, which was so characteristic of the great French Revolution, when the French Convention launched into sweeping measures but did not possess the necessary base of support in order to put them into effect, and did not even know on what class to rely in order to put any particular measure into effect.

Our position is an infinitely more fortunate one. Thanks to a whole century of development, we know on which class we are relying. But we also know that the practical experience of that class is extremely inadequate. The fundamental aim was clear to the working class and the workers' party—to overthrow the power of the bourgeoisie and to transfer power to the workers. But *how* was that to be done? Everyone remembers with what difficulty and at the cost of how many mistakes we passed from workers' control to workers' management of industry. And yet that was work within our own class, within the proletarian midst, with which we had always had to deal. But now we are called upon to define our attitude towards a new class, a class the urban worker does not know. We have to determine our attitude towards a class which has no definite and stable position. The proletariat in its mass is in favour of socialism, the bourgeoisie in their mass are opposed to socialism. It is easy to determine the relations between these two classes. But when we pass to a section like the middle peasants we find that *it is a class that vacillates.* The middle peasant is partly a property-owner and partly a working man. He does not exploit other working people. For decades the middle peasant defended his position with the greatest difficulty, he suffered the exploitation of the landowners and the capitalists, he bore everything. Yet he is a property-owner. Our attitude towards this vacillating class therefore presents enormous difficulties. In the light of more than a year's experience, in the light of more than six months' proletarian work in the rural districts, and in the light of the fact that class differentiation in the rural districts has already taken place, we must most of all beware here lest we are too hasty, lest we are inadequately theoretical, lest we regard what is in process of being accomplished, but has not yet been realised, as already accomplished. In the resolution which is being proposed to you by the commission elected by the committee, and which will be read to you by a subsequent speaker, you will find sufficient warning against this.

From the economic point of view, it is obvious that we must help the middle peasant. Theoretically, there is no doubt of this. But because of our habits, our level of culture, the inadequacy of the cultural and technical forces we are in a position to place at the disposal of the rural districts, and because of the impotent manner in which we often approach the rural districts, comrades quite often resort to coercion and thus spoil everything. Only yesterday a comrade gave me a pamphlet entitled *Instructions and Regulations on Party Work in Nizhni-Novgorod Gubernia,* issued by the Nizhni-Novgorod Committee of the Russian Communist Party (Bolsheviks), and in this pamphlet, for example, I find on p. 41: "The whole burden of the extraordinary tax decree must be placed on the shoulders of the village kulaks and profiteers and *the middle element of the peasants generally.*" Well, well! These people have indeed "understood". This is either a printer's error—and it is impermissible that such printer's errors should be made—or a piece of rushed, hasty work, which shows how dangerous all haste is in this matter. Or—and this is the worst surmise of all, one I would not like to make with regard to the Nizhni-Novgorod comrades—they have simply failed to understand. It may very well be that it is an oversight.

We have, in practice, cases like the one related by a comrade in the commission. He was surrounded by peasants, and every one of them asked: "Tell me, am I a middle peasant or not? I have two horses and one cow. I have two cows and one horse," etc. And this agitator, who tours the uyezds, is expected to possess an infallible thermometer with which to gauge every peasant and say whether, he is a middle peasant or not. To do that you must know the whole history of the given peasant's farm, his relation to higher and lower groups—and we cannot know that accurately.

Considerable practical ability and knowledge of local conditions are required here. And we do not possess this yet. You need not be ashamed to confess it; it must be admitted frankly. We were never utopians and never imagined that we would build communist society with the immaculate hands of immaculate communists, born and educated in an immaculately communist society. That is a fairy-tale. We have to build communism from the debris of capitalism, and only the class which has been steeled in the struggle against capitalism can do that. The proletariat, as you are very well aware, is not free from the shortcomings and weaknesses of capitalist society. It is fighting for socialism, but at the same time it is fighting its own shortcomings. The best and foremost section of the proletariat, which carried on a desperate struggle in the cities for decades, was in a position to acquire in the course of that struggle the urban culture and that of life in the capital; and to a certain extent it did acquire it. You know that even in advanced countries the rural districts were condemned to ignorance. Of course, we shall raise the level of culture in the rural districts, but that will be a work of many, many years. That is what our comrades everywhere are forgetting and what is being strikingly brought home to us by every word uttered by people who come from the rural districts; not by the intellectuals who work here, not by the officials—we have listened to them a lot—but by people who have in practice observed the work in the rural districts. It was these opinions that we found particularly valuable in the agrarian committee. These opinions will be particularly valuable now—I am convinced of that—for the whole Party Congress, for they come not from books, and not from decrees, but from experience.

All this obliges us to work for the purpose of introducing the greatest possible clarity into our attitude towards the middle peasant. This is very difficult, because *such clarity does not exist in reality.* Not only is this problem unsolved, it is unsolvable, if you want to solve it *immediately and all at once.* There are people who say: "There was no need to write so many decrees." They blame the Soviet Government for setting about writing decrees without knowing how they were to be put into effect. These people, as a matter of fact, do not realise that they are sinking to the whiteguard position. If we had expected that life in the rural districts could be completely changed by writing hundreds of decrees, we would have been absolute idiots. But if we had refrained from indicating in decrees the road that must be followed, we would have been traitors to socialism. These decrees, while in practice they could not be carried into effect fully and immediately, played an important part for propaganda. While formerly we carried on our propaganda by means of general truths, *we are now carrying on our propaganda by our work.* That is also preaching, but it is preaching by action—only not action in the sense of isolated sallies of some upstarts, at which we scoffed so much in the era of the anarchists and the socialism of the old type. Our decree is a call, but not the old call: "Workers, arise and overthrow the bourgeoisie!" No, it is a call to the people, it calls them to practical work. *Decrees are instructions which call for practical work on a mass scale.* That is what is important. Let us assume that decrees do contain much that is useless, much that in practice cannot be put into effect; but they contain material for practical action, and the purpose of a decree is to teach practical steps to the hundreds, thousands, and millions of people who heed the voice of the Soviet government. This is a trial in practical action in the sphere of socialist construction in the rural districts. If we treat matters in this way we shall acquire a good deal from the sum total of our laws, decrees, and ordinances. We shall not regard them as absolute injunctions which must be put into effect instantly and at all costs.

We must avoid everything that in practice may tend to encourage individual abuses. In places careerists and adventurers have attached themselves to us like leeches, people who call themselves

Communists and are deceiving us, and who have wormed their way into our ranks because the Communists are now in power, and because the more honest government employees refused to come and work with us on account of their retrograde ideas, while careerists have no ideas, and no honesty. These people, whose only aim is to make a career, resort in the localities to coercion, and imagine they are doing a good thing. But in fact the result of this at times is that the peasants say: "Long live Soviet power, but *down with the communia!*" (i.e., communism). This is not an invention; these facts are taken from real life, from the reports of comrades in the localities. We must not forget what enormous damage is always caused by lack of moderation, by all rashness, and haste.

We had to hurry and, by taking a desperate leap, to get out at any cost of the imperialist war, which had brought us to the verge of collapse. We had to make most desperate efforts to crush the bourgeoisie and the forces that were threatening to crush us. All this was necessary, without this we could not have triumphed. But if we were to act in the same way towards the middle peasant it would be such idiocy, such stupidity, it would be so ruinous to our cause, that only provocateurs could deliberately act in such a way. The aim here must be an entirely different one. Here our aim is not to smash the resistance of obvious exploiters, to defeat and overthrow them—which was the aim we previously set ourselves. No, now that this main purpose has been accomplished, more complicated problems arise. You cannot create anything here by coercion. *Coercion applied to the middle peasants would cause untold harm.* This section is a numerous one, it consists of millions of individuals. Even in Europe, where it nowhere achieves such strength, where technology and culture, urban life and railways are tremendously developed, and where it would be easiest of all to think of such a thing, nobody, not even the most revolutionary of socialists, has ever proposed adopting measures of coercion towards the middle peasant.

When we were taking power we relied on the support of the peasants as a whole. At that time the aim of all the peasants was the same—to fight the landowner. But their prejudice against large-scale farming has remained to this day. The peasant thinks: "A large farm, that means I shall again be a farm-hand." That, of course, is a mistake. But the peasant's idea of large-scale farming is associated with a feeling of hatred and the memory of how the landowners used to oppress the people. That feeling still remains, it has not yet died.

We must particularly stress the truth that here, by the very nature of the case, coercive methods can accomplish nothing. The economic task here is an entirely different one. Here there is no upper layer that can be cut off, leaving the foundation and the building intact. That upper layer which in the cities was represented by the capitalists does not exist here. *Here coercion would ruin the whole cause.* What is required here is prolonged educational work. We have to give the peasant, who not only in our country but all over the world is a practical man and a realist, concrete examples to prove that the "communia" is the best possible thing. Of course, nothing will come of it if hasty individuals flit down to a village from a city, come there, chat about, stir up a number of intellectual-like and at times unintellectual-like squabbles, and then quarrel with everyone and go their way. That sometimes happens. Instead of evoking respect, they evoke ridicule, and deservedly so.

On this question we must say that we do encourage communes, but they must be so organised *as to gain the confidence of the peasants.* And until then we are pupils of the peasants and not their teachers. Nothing is more stupid than when people who know nothing about agriculture and its specific features, people who rush to the village only because they have heard of the advantages of socialised farming, are tired of urban life and desire to work in rural districts—when such people regard themselves as teachers of the peasants in every respect. *Nothing is more stupid than the very idea of applying coercion in economic relations with the middle peasant.*

The aim here is not to expropriate the middle peasant but to bear in mind the specific conditions in which the peasant lives, to learn from the peasant methods of transition to a better system, *and not to dare to give orders!* That is the rule we have set ourselves. [*General applause*] That is the rule we have endeavoured to set forth in our draft resolution, for in that respect, comrades, we have indeed sinned a great deal. We are by no means ashamed to confess it. We were inexperienced. Our very struggle against the exploiters was taken from experience. If we have sometimes been condemned on account of it, we can say: "Dear capitalist gentlemen, you have only yourselves to blame. If you had not offered such savage, senseless, insolent, and desperate resistance, if you had not joined in an alliance with the world bourgeoisie, the revolution would have assumed more peaceful forms." Now that we have repulsed the savage onslaught on all sides, we can change to other methods, because we are acting not as a narrow circle, but as a party which is leading the millions. The millions cannot immediately understand a change of course, and so it frequently happens that blows aimed at the kulaks fall on the middle peasants. That is not surprising. It must only be understood that this is due to historical conditions which have now been outlived and that the new conditions and the new tasks in relation to this class demand a new psychology.

Our decrees on peasant farming are in the main correct. We have no grounds for renouncing a single one of them, or for regretting a single one of them. But if the decrees are right, *it is wrong to impose them on the peasants by force.* That is not contained in a single decree. They are right inasmuch as they indicate the roads to follow, inasmuch as they call to practical measures. When we say, "encourage associations", we are giving instructions which must be tested many times before the final *form* in which to put them into effect is found. When it is stated that we must strive to gain the peasants' voluntary consent, it means that they must be persuaded, and persuaded by practical deeds. They will not allow themselves to be convinced by mere words, and they are perfectly right in that. It would be a bad thing if they allowed themselves to be convinced merely by reading decrees and agitational leaflets. If it were possible to reshape economic life in this way, such reshaping would not be worth a brass farthing. It must first be proved that such association is better, people must be united in such a way that they become actually united and are not at odds with each other—it must be proved that association is advantageous. That is the way the peasant puts the question, and that is the way our decrees put it. If we have not been able to achieve that so far, there is nothing to be ashamed of and we must admit it frankly.

We have so far accomplished only the fundamental task of every socialist revolution—that of defeating the bourgeoisie. That in the main has been accomplished, although an extremely difficult half-year is beginning in which the imperialists of the world are making a last attempt to crush us. We can now say without in the least exaggerating that *they themselves understand that after this half-year their cause will be absolutely hopeless.* Either they take advantage now of our state of exhaustion and defeat us, an isolated country, or we emerge victorious not merely in regard to our country alone. In this half-year, in which the food crisis has been aggravated by a transport crisis, and in which the imperialist powers are endeavouring to attack us on several fronts, our situation is extremely difficult. But *this is the last difficult half-year.* We must continue to mobilise all our forces in the struggle against the external enemy, who is attacking us.

But when we speak of the aims of our work in the rural districts, in spite of all the difficulties, and in spite of the fact that our experience has been wholly concerned with the immediate task of crushing the exploiters, we must remember, and never forget, that our aims in the rural districts, in relation to the middle peasant, are entirely different.

All the class-conscious workers—from Petrograd, Ivanovo-Voznesensk, or Moscow—who have been to the rural districts related examples of how a number of misunderstandings which appeared to be irremovable, and a number of conflicts which appeared to be very serious, were removed or mitigated when intelligent working men came forward and spoke, not in the bookish language, but in a language understood by the peasants, when they spoke, not as commanders who take the liberty of giving orders without knowing anything of rural life, but as comrades, explaining the situation and appealing to their sentiments as working people against the exploiters. And by such comradely explanation they accomplished what could not be accomplished by hundreds of others who conducted themselves like commanders and superiors.

That is the spirit that permeates the resolution we are now submitting to your attention.

I have endeavoured in my brief report to dwell on the underlying principles, on the general political significance of this resolution. I have endeavoured to show—and I should like to think that I have succeeded—that from the point of view of the interests of the revolution as a whole we are making no change of policy, we are not changing the line. The whiteguards and their henchmen are shouting, or will shout, that we are. Let them shout. We do not care. We are developing our aims in a most consistent manner. We must transfer our attention from the aim of suppressing the bourgeoisie to the aim of arranging the life of the middle peasant. We must live in peace with him. In a communist society the middle peasants will be on our side only when we alleviate and ameliorate their economic conditions. If tomorrow we could supply one hundred thousand first-class tractors, provide them with fuel, provide them with drivers—you know very well that this at present is sheer fantasy—the middle peasant would say: "I am for the communia" (i.e., for communism). But in order to do that we must first defeat the international bourgeoisie, we must compel them to give us these tractors, or so develop our productive forces as to be able to provide them ourselves. That is the only correct way to pose this question.

The peasant needs the industry of the towns; he cannot live without it, and it is in our hands. If we set about the task properly, the peasant will be grateful to us for bringing him these products, these implements, and this culture from the towns. They will be brought to him, not by exploiters, not by landowners, but by his fellow-toilers, whom he values very highly, but values in a practical manner, for the actual help they give, at the same time rejecting—and quite rightly rejecting—all domineering and "orders" from above.

First help, and then endeavour to win confidence. If you set about this task correctly, if every step taken by every one of our groups in the uyezds, the volosts, the food detachments, and in every other organisation is made properly, if every step of ours is carefully checked from this point of view, we shall gain the confidence of the peasant, and only then shall we be able to proceed farther. What we must now do is to help him and advise him. This will not be the orders of a commander, but the advice of a comrade. The peasant will then be entirely on our side.

This, comrades, is what is contained in our resolution, and this, in my opinion, must become the decision of the Congress. If we adopt this, if it serves to determine the work of all our Party organisations, we shall cope with the second great task confronting us.

We have learnt how to overthrow the bourgeoisie, how to suppress them, and we are proud of the fact. But how to regulate our relations with the millions of middle peasants, in what way to win their confidence, that we have not yet learnt—and we must frankly admit it. But we have understood the task, we have set it, and we say in all confidence, with full knowledge and determination, that we shall cope with this task—and then socialism will be absolutely invincible. [*Prolonged applause*]

CUBA'S AGRARIAN REFORM
by Fidel Castro

Delegates: To our memory comes the history of the whole revolutionary process in our agriculture. This Congress which has just taken place has great importance for the Revolution, but perhaps its importance is not as obvious today as it will be in the years to come. I believe that this Congress represents a great step forward.

But it is necessary that all of you compañeros, who work in the cane fields, understand clearly why this signifies a great step forward. First of all, we ought to explain the first steps that were taken in agriculture, and why they were taken.

All of you who work in the cane plantations have a very clear idea of what life was like in the countryside, above all on the cane plantations.

When the Revolution triumphed, it was a fact that the first step taken was the Agrarian Reform. You will remember when we began to speak of the Agrarian Reform, how immediately the people began to be interested, even the workers in the cities.

It is possible that many people heard of the Agrarian Reform, without, nevertheless, understanding what this signified. But finally everyone realized that the Reform was necessary, that the situation in the countryside could not remain, that a complete change was needed in the conditions of life, work and exploitation of the land, that such a change would be beneficial to the campesinos (farmers).

The Agrarian Reform is one of the most completed tasks which the Revolution set itself. It has also been one of the most difficult tasks. The problem of ownership and means of exploitation of the land is far more complex than the same problem in industry. The revolution, for example, of the system of industrial production is always much simpler than the revolution in the countryside.

In our countryside, two types of production centres existed: the large latifundios (plantations) and the small farmers. The large latifundios exploited a considerable number of workers, especially the large cane latifundios. Among the small farmers there were certain distinct characteristics. The small farmer who was the owner of land was in a minority. Then there were the squatters, the cultivators of coffee and cocoa in the mountains, who, although they did not pay rent, always lived under the threat of eviction. There was also the farmer who paid rent, who, together with the squatters, constituted the great majority.

We were, therefore, confronted with two types of ownership on the land. There was the small farmer who worked his own land and there was the landlord who lived, in many cases, far from his lands and who employed at times hundreds of workers. The large latifundios like the United Fruit Co., employed thousands of workers.

The basic outlines of the first revolutionary law that changed the system of production in the countryside were as follows: first, the liquidation of the latifundios; second, the liquidation of the rent system, that is to say the liberation of the campesino from the rents that they were paying; third, to guarantee to the squatters the ownership of the land that they were occupying.

The problem remained of how to make the great latifundios productive. It was talked about a great deal, and it was on everybody's tongue in a period when we could aspire only to partial triumphs, to

---

Speech at the closing session of the National Congress of Cane Cooperatives, August 18, 1962.

partial solutions. It was a period when we could not move forward with the Revolution on all fronts, but had to move in stages. In this period, much was said of the Agrarian Reform as simply a distribution of the lands. Many people saw the Agrarian Reform as nothing more than the distribution of land.

Fortunately, our revolution had the good judgment and sufficient audacity to adopt a progressive system of exploitation of the land. Today that is very easy to understand.

Why we didn't divide the latifundios
The division of the latifundios could have destroyed the Revolution. Dramatic problems would have resulted from the division of these lands. First of all there is the practical problem of dividing these lands since all the lands have distinct characteristics. Within any latifundio some land is more fertile than other, some is used for one thing, some for another. From the political point of view, the easiest solution would have been the division of those lands. From the practical point of view, this would have been an inferior solution. Often, that which at one moment appears easiest is in the long run not the best.

The results that would have followed the division of these lands, are understood by everyone. In the first place, there isn't enough land for everyone. We remember that when the agrarian law was proclaimed, someone suggested printed forms for those who wanted land. We realized then that such forms would have negative results. Why? Because everyone wanted land, even the people who lived in the cities.

If the latifundios had been divided, many workers would have been left without land or, on the other hand, the lots would have been too small for a family to subsist on.

Can you imagine, for example, a rice latifundio of one or two hundred caballerías planted in rice, divided among three or four hundred families? Naturally, each would take possession of his parcel of land and would build his house there. Besides rice, he would wish to grow food and many other things. In any rice latifundio the system of irrigation requires the flooding of up to 30 caballerías, (one caballería = 33 acres) therefore we would have remained isolated like an island reef in the middle of flooded rice fields.

With sugar cane, more or less the same thing would have happened. With cattle it would have been even worse. The distribution of the cattle and the cattle ranches would have created one of the most serious problems of the Revolution.

First of all, it is hard to imagine the number of cattle that would have been sacrificed, especially with the present increase in the demand for meat. At the same time the demand for shoes has also increased considerably, and therefore an increase in the demand for leather.

We do not have the problem solved as yet, but we have created all the conditions for the resolution of these problems. Today there is considerably less meat than the people wish to buy. It is obvious that to satisfy this demand, we would have to sacrifice not only the fattened cattle, but also the cows and even the heifers and calves. In a word, if we were to kill all the cattle that the people today are able to buy, then in three or four years very few cattle would remain.

Another problem as a result of this policy would be a reduction in the number of hides available for the manufacture of footwear. You know about the difficulties in the production of footwear; nevertheless, only increased production will solve this problem. In this case the production of meat from large animals is tied closely to the production of footwear.

Imagine the situation if we were to dispose of half the cattle we have, and could not say like today, that year by year, we can kill so many cattle, which will supply so much leather for so many pairs of shoes.

But that is not all: We have a program for the development of the cattle herds through artificial insemination and through importation and selection of breeding stock.

To start a program of insemination and selection in a large centre where there are hundreds of thousands of cattle does not represent the same problems that would arise if there were tens or hundreds of thousands of small producers developing, for example, the beef herds. Also the breeding of hogs requires certain conditions; veterinarians, feeding houses, etc. It would be almost impossible to carry out a plan of rapid development where the land was divided into tens of thousands of farms, each with its own style, methods and ideas. It would be the same for any program of selection, of cultivation improvement, of irrigation. That is to say, production problems would represent an insurmountable situation.

We are not going to talk about other problems which are related to the agricultural workers, that is, to those who live in the country. If the latifundios had been divided, each one would have built his own bohio on his own little piece of land; the school would have remained several kilometers from where the children live; the possibilities of electrification, suitable roads, sewage, recreation sites and shopping centres would not have been realised.

None of the many towns that have been built, that nevertheless are only a small part of what are really needed in the countryside, could have been realised either. The possibility of bringing the comforts of city life to the countryside would never have been possible, i.e., the large school centers, artistic and cultural activities, running water, electricity, sewage, streets; in a word, everything that is done in a small town would have been impossible in the countryside since such small towns can be built only when the land is cultivated collectively.

Besides this, the division of the land would have had other economic, political and social consequences. It is a fact that today there is a great demand for products, which, as a result of drought, poor administration and errors, agricultural production did not increase in proportion to the demand. We see, as a result, some speculation in agricultural products and more or less exorbitant prices.

Where does it occur and how does it happen? Why? Why does meat keep its price? Why do fowl, which are distributed in the cities, keep their price? Why do those staple articles not vary in price? Why do pork products keep their price? Why aren't three turkeys worth 50 pesos? Why don't they sell four chickens for 20 pesos in the cities? Why? Because those products come from the Granjas (Peoples Farms or State Farms). Because the beef which is used in the cities was raised on the Granjas! Because those fowl were raised on the Granjas! Because the agricultural products of the Granjas come to the towns and cities, to the workers, at reasonable and stable prices.

The Nation can count on those products because they go from the production centers to the distribution centers. What are the products which become objects of speculation? They are the products cultivated on the small isolated plots.

What happened, for example, to the malanga from "Rancho Mundito", in Pinar del Rio, malanga which was sown with loans given by the Revolution, on land which was distributed by the Revolution? The campesinos made use of the credits; they sowed and they produced, but what happened then? The Revolutionary Government, in order to avoid annoyances for the independent producers, authorized the campesinos to sell on the open market. What happened? Everyone with a car and money went to "Rancho Mundito" to buy malanga; some for their own use, others for purposes of speculation.

The result was that the area which supplied malanga for the children of the city of Havana, sold in a single Sunday 300,000 lbs. of malanga, directly, to the people from the city; that is, enough malanga for the children of Havana for several days.

And what was the result of this? The person who went with his car and bought a hundred weight of malanga, was assured of the supply for his children, his own children, for several weeks. The speculator, who gained by reselling, forced families, who could find no malanga on the store shelves, to pay three or four times the regular price. And the children of families who had no car and who could not afford to pay three or four times the regular price had to go without.

In the face of shortages, the speculator was always present, offering much higher prices and creating difficulties for the whole population. What is even worse, the speculator was corrupting the campesino, awakening excessive ambitions, ambitions which they are not able to satisfy without causing suffering for the rest of the citizens. No individual in society can disregard the needs of others; no one can disregard others.

Anyone, then, who lives at the expense of others is acting in an anti-social and unjust way. He forgets that he needs others because the one who produces a specific crop and wishes to sell for ten times the proper price, would not like, for example, to smoke a cigar that cost him ten times more than the regular price. He would not like to pay ten times more for shoes and clothing than is just; nor would he like it if the kerosene or the medicine which he gets or the rice, beans, salt, sugar, the thousand products which he needs to live, cost him ten times more than they should. If we forget this, we forget, in short, that we all work for others and we all need the work of others.

Who benefits from speculation?

So we see that within society articles of every type are needed—how even here at this gathering, we need electric lights and the chairs on which you are sitting. There were workers, carpenters, decorators, who organized all this. The clothes that you wear, the shoes, the room where you sleep, the meals they serve you, the trains in which you travel: they are products of the work of others. We would not have light; we would not have telephones; we would not have transportation, gasoline, clothing or shoes if each one kept what he produced. We would have no medicine, no teachers, no doctors, no mail. We would have practically nothing because it is an elemental truth that we are all dependent on the work of others. If someone receives ten times the just price for something he produces, then he is stealing from others.

Anyone who wishes to sell an article for ten times what it costs to produce, and to buy what he uses for a just price, cannot do so without robbing the rest of the citizens. And when a bourgeois from the city comes in his automobile and offers a campensino for a hundred pounds of malanga, not $4, which is the proper price, more or less, a price which is in line with what other workers can pay for the article, but offers him $10 for the hundred weight, the bourgeois corrupts the campesino with a price three or four times more than he thought was possible. He does this calmly, like a person doing the right thing, because he has the right to sell that food for $10 or even $15; or he goes to the highway and sells three turkeys for $50.

The campesino does not understand these things. In his lifetime he has been concerned only with the accumulation of money, because he lived in a society where money was everything. And who are the ones who eat this high priced food? Who are the ones who eat fried turkey? The rich! The bourgeois in the cities! [*Applause*]

When an 80 lb. pig sells for $80, what worker can afford to buy pork? The worker who earns only $80 or $100 per month, so that sugar, salt and meat can be produced cheaply for the whole Nation, that worker cannot eat pork or turkey, his children cannot eat malange simply because only the ones who have cars and who have large incomes or profits, and who besides, produce absolutely nothing, can go to the countryside.

I think that everyone here understands these things, especially you workers who have worked all your lives! You understand these things perfectly! [*Applause*]

The result! The rich who remain, live well; but this is not the worst. They sow greed, ambition, corruption and demoralization among the small farmers. The small farmer is a humble man who is a worker, not a parasite. But the parasite cannot be a parasite if, along the way, he is not fulfilling acts of parasitism, sowing parasitism and creating parasites where ever he goes. He is not satisfied with being a parasite; he must convert the campesino into a parasite also. [*Applause*]

All this does not happen on a Granja. With all the deficiencies, with all the errors, with all the things badly done, this does not happen on a Granja. If there are caballerías of malanga, no bourgeois is allowed to come and offer $10 a hundred weight. [*Applause*] No speculator can go there to buy at a higher price in order afterwards to rob the workers. The people can count on the 10 caballerías there; all the children can count on a just price. [*Applause*]

In the same way, the worker of that farm has the right to buy industrial products at a just price: cigars, tobacco, all the food which is not produced on that farm; clothing, shoes, medicines, transportation, everything is sold to him at a fair price.

I have enumerated, therefore some distinct consequences. If the latifundios had been divided, not only would production have dropped in a significant way, but we would have been left without a base to develop our agricultural economy rapidly. We would not have been able to reduce unemployment in the country. Not only that; the staple commodities would have been in short supply and speculation would have been rampant.

Another thing is also very clear: It is common for a Granja in the Havana area to produce, with hundreds of workers, 1,000 or 2,000 litres of milk. We can say with all certainty that if this milk were sold on the open market, then the people of Havana would not be assured of a supply of milk. But all this milk should not be consumed on the farm; the remainder must be used by the rest of the population, by the workers and their families who live in the cities. All the milk is not consumed on the farm; part of it is distributed and the farm worker reasons "that milk is lacking in the city, we cannot leave the city without milk." This could not be done if the latifundios had been divided, because the private producer first consumes all that he needs. After he has satisfied his own needs he sells the surplus to the highest bidder, because—imagine—he still does not understand his social obligations; the products are in their hands.

I do not wish to go into detail at this time, for example, of the problem of sending thousands of young people to study in the Soviet Union. It is much easier to make arrangements with Granjas or Co-operatives than with 600 individual farmers. [*Applause*]

Rural unemployment liquidated

From every point of view, and today we see it with total clarity, it was a great thing, a great step, when the Revolution turned the latifundios into collective centres of production; in spite of all the difficulties, all the deficiencies at every turn, it was a great step.

You know that rural unemployment has been liquidated. You know that the problem now is a shortage of labor in many parts of the country. [*Applause*] You know now all that, know that you no longer have to seize the "matulito," the "saquito" (small bag containing earthly belongings), carrying in your arms and taking by the hand your starving children—to go picking coffee, [*Applause*] or to emigrate to far away places, or to look for a political boss who could hand out a public works job, or who would buy your electoral ballot to be cast for some shameless politician. [*Applause*] Now it is not necessary to do that.

For that reason, those campesinos who, like yourselves, came from the latifundios, went as squatters into the precarious mountains to sow coffee. The coffee was cultivated in the mountains because the cane and cattle latifundios devoured all the good land, and were not interested in coffee. If there is coffee in Cuba today it is thanks to those campensinos, who, during the "dead period" (when there was no work in the cane fields) when thousands of workers were hungry, scaled the mountains: thanks to them there is coffee.

This coffee, which was in remote places, could be picked because there were tens and hundreds of thousands of men in the countryside who were out of work.

When the Revolution came and resolved the fundamental problem of unemployment in the countryside, resolved it because it neither distributed nor divided those latifundios, thereby permitting the mechanization of agriculture, the Revolution now has another problem: who picks the coffee?

But the Revolution finds solutions for everything, [*Applause*] because the Revolution by its own deeds draws upon fresh forces; by its own deeds finds new resources and like the workings of the Revolution, itself is an extraordinary development. Tens and tens of thousands of young people are studying, their entire expenses are paid by the Nation. The Revolution mobilizes these young people, mobilizes their energy—the same as it mobilized them when they went to teach in the mountains—and resolves the problems with their resources and group energy. [*Applause*]

For this reason, the campesinos are not left now with no one to pick their coffee. They don't have to worry because their brothers have now work on the prairies, because the Revolution finds solutions for the problems of all. The campesino should not complain about the good fortune of his brothers, because he *will* have someone to pick the coffee.

Thus we see that despite the many shortages, and they are many, something has been done, changes have been made. As we look back on our difficulties and errors, we must apply new means. We are not yet beginning to ascend, but already we have walked part way up hill. We do not now have the problems of the first days, the problems there were in the beginning. There are problems, yes, but problems that correspond to a stage where we have overcome many of the past evils.

Co-operatives and state farms

How did we go about organizing production in the countryside? What did we do? The latifundios were not divided, but collective centers of production were established. Two types appeared; the co-operatives in the cane fields, and Granjas in the cattle and rice latifundios, and on the virgin lands.

The Revolution took a bold step when it did not distribute the latifundios. The two types of collective agricultural organization, the co-operative and the Granja, have been marching side by side.

From the cane latifundios, more than 600 Cane Cooperatives have been organized. From the cattle latifundios and virgin lands, more than 300 Granjas have been organized or constructed.

You all know, compañeros, how much interest we have shown, the meetings that have been held, the courage we have demonstrated in trying to advance the co-operatives. Who here is not aware of the initiative that was shown in placing a dairy in each co-operative, the credits that were given, the improvements that were planned, the towns, the solution of the housing problem?

Naturally, many of these things were impossible to solve in such a brief space of time, such as the problem of housing.

Many courses have been organized to produce mechanics, agricultural technicians. You know that there is not a single co-operative where many young people have not gone away to study.

These two parallel systems have been subjected to the test of reality. The co-operative is a collective centre distinct from the Granja: the Granja is like a factory, the farmer is like a worker in a factory; the co-operative worker is like a member of a group of workers who works for his own benefit and not for the benefit of the Nation.

It is logical that the bookkeeping of the one type would be different from the other. If the co-operative worker works for his own benefit, then he receives only the land free; not the investments. The investment, the machinery must be paid for. Housing must also be his own responsibility. If the product is going to be his, the instruments of production, the investments, the housing, all must be paid for.

The case of the worker on the Granja is different. In the first place, the machinery, the investments, do not have to be paid for by him. But there is something more: the Revolution decided that housing would not have to be paid for either, nor electricity and water.

Now we see the difference. Because the worker on the Granja works for the benefit of the Nation, he has the right to receive the instruments of production as well as all the possible benefits that the Nation can give. Not so with the co-operative worker: for him there must be a charge.

During the initial stages of building the co-operative centres, therefore, credit was required for the construction of the towns. Many years of work will be required to pay for the investments, the machinery, the housing. Naturally, if this is not accompanied by high productivity, and if the people live well in the meantime, then many years will be required to pay for everything. How long, it cannot be known precisely.

It was logical also, on the other hand, that if the state imported 10,000 head of milk-producing cattle, they would not take them to the co-operative, but would take them rather to the Granjas, where the products are for the whole Nation. If we import 20,000 hogs, they too must be taken to the Granjas. Equipment for artificial insemination of cattle and special seeds must first be taken to the Granjas. If we introduce new techniques, such as hybrid corn, they must first go to the Granjas. We could even say that the workers on the Granjas should be the first to get houses, because they work for the Nation as a whole while the co-operative worker works for his own benefit. [*Applause*]

But there was also another problem which we did not have on the Granja. Those who worked on the Granjas were workers who did not exploit anyone. Everyone was equal. But on the co-operatives, a problem existed, where a determined number of people were co-operativistas (members of the co-operative who shared in the profits). Others, what were they? Second class citizen-workers; marginal ones. They were nothing. Since they were not co-operativistas they were workers and the co-operativistas were employers. [*Applause*] When profits were distributed, they received nothing. The co-operativistas got the first houses and other advantages.

Since agriculture requires more labor at certain times of the year than at others, it is sad to think that the rural worker now is similar to the martyrs of the pre-revolutionary period of exploitation. The Revolution came, and although things are much better now than what they had been, nevertheless, this situation is hard to accept with resignation.

There remained in the countryside, an outcast,

who was not a co-operativista, who was nothing. Why? Because he had that shame, he was different from others. Nevertheless his needs were the same as the others, the same preoccupations, therefore he ought to have the same rights. [*Applause*]

It is true that although a great step forward had been made, that forward step did not correspond entirely to the Revolution's idea of justice.

Who had suddenly become, at times, a semi-exploiter of the work of others? Old workers! A contradiction exists here: the latifundio has been liquidated and this is a great forward step. But at the same time a group of proletarians, one of the most combative, most inured to war and most revolutionary—as was the cane worker—have lost their position as proletarians. The Revolution was taking a backward step.

And where were the agricultural workers who were the most revolutionary, the most long suffering, the best fighters? They were not on the cattle latifundios, managed by a few laborers. Traditionally, the group with the best fighting spirit, the most revolutionary of the agricultural proletariat, was the cane worker, the workers of the cane latifundios. When the Revolution was becoming more proletarian; when the country moved forward to the great moment when its destiny was not in the hands of the bourgeois exploiters, nor of the imperialist filibusters; at the moment when the proletariat began to guide the destiny of the country, a great proletarian and exploited group of yesterday ceased being proletarian. [*Applause*]

For you, compañeros—although the co-operative was a forward step over the past, more just than the latifundios, a forward step for the Nation—for you, from the point of view of your class, for you from the point of view of historical importance, taking into account the class that you belong to, for you the co-operative was a backward step.

I am sure that in the mind of each one of you is indelibly engraved all of the past—the past—of leaders, of rural guards, of unscrupulous politicians, of gamblers, of boliteros (private lottery owners), of grafters of every type. You who have deeply engraved in your mind the memory of that hungry past, the past of suffering and humiliated workers, you would not have thought that the day would come when your children would study in university or in foreign lands. How could a worker have thought of becoming an administrator, of coming to the capital to discuss problems, those who are sent by their place of work, [*Applause*] to discuss with the Ministers, with the Government, to express their opinions, to participate actively in the affairs of the country? You who remember that past, I am sure, if you were asked to renounce your position as proletarians to become semi-exploiters, I am sure that you would all say: "No, we will not renounce our proletarian position. We now, more than ever, wish to be proletarians, because in our hands is the destiny of our country. We wish to change it into a better world, without exploiters or exploited of any kind." [*Applause*]

That, compañeros, is the way we must think. To be a proletarian is an honour and title above any others in our society; more than material advantages you may be able to acquire.

Yesterday it was the owner of the latifundio; yesterday the Yankee was the boss and made the "rendez-vous" every day. Today, the highest honour, the master of our country, is not the Yankee, nor is it the landlord exploiter. It is the proletarian! [*Applause*]

The co-operatives have fulfilled their usefulness. They had deficiencies, great deficiencies. Each of the two types of production has been tested. Is it not correct to take one more forward step, a step that will bring together all the workers of the countryside, the workers of the Granjas and you (co-operativistas), bringing together you and those who work with you in the cane fields but who are not co-operativistas?

Agricultural working class advance

With this step, the agricultural proletariat begins to advance, becomes the most numerous sector of the working class in our country—great in size, number and importance—because altogether, on the co-operatives and on the Granjas, there are more than 250,000. With this step, the Revolution will have 250,000 rural proletarians, [*Applause*] a great and formidable force for the Revolution.

When the small independent farmer unites in order to produce more economically, he advances. For that reason, the true co-operative is one which is formed from independent producers who are not proletarians. [*Applause*] The small farmer is the one who is attached to the land, to his small plot of ground. He has a feeling of ownership that the proletariat does not have. That (the land) is his world; he does not have the advanced mentality of the proletariat. When he joins with others, that is a forward step for him and for the Revolution.

Nearly 300 Agricultural Societies of Farmers, have been created, resulting in the joining of lands. [*Applause*]

Now this is a very complex problem, because the campesino does not have the same mentality as he who has been, or is, a worker. The campesino has a different outlook: he does not have the same level of culture, above all of political consciousness, which the worker has. It is necessary to march with the campesino, who is an ally of the working class. We must go to the campesino, each time raising his consciousness, each time making him more revolutionary, each time more advanced. For this, it is necessary to have a correct policy. With proper leadership, the campesino will move spontaneously toward superior forms of production. The campesino cannot be socialized or co-operativized by force or compulsion.

Neither by force nor compulsion
No, that campesino must be allowed to develop little by little until, of necessity, there will no longer be any workers looking for a boss. When the Granjas encompass all labour, then the yoke of oxen will not do. It will be necessary to mechanize everything; then the campesino will see that by uniting forces with other campesinos he will have more strength, more possibilities and will be able to produce more. Then he will march forward.

The day will come when no one will want to work for a boss. [*Applause*] That day will arrive as it is now arriving, as it is now a reality in many places, when all farmers will move toward the Granjas, to remunerative work and to all the benefits of the Granja. Not the Granja of today, because we still lack an infinite number of things. Now there are many problems to solve, even such elemental ones as housing. No, the centres which are now being created, the communities, the towns which the campesinos will have, the life in these communities will, in every way, be comparable to life in the cities.

Here, today, some agricultural workers in this theatre (where the Congress was being held), entertained us with their songs and with their artistic achievements. Thousands of instructors are being trained now to make these possibilities a reality on the Granjas, and in the countryside, where life is isolated, tedious, often boring. They will organize the life of the future where highly productive work will alternate with many other pleasant things which will make the life of our campesinos much better: a healthy, happy life, a life of work, healthy diversions, sports and recreation.

The day will come when lamps will no longer be used; the day will come when many of the things that are seen now only in the cities, will be seen on the farms as well. And that day is not so far off, not so far that you will not see it. [*Applause*]

That is what we must think about, because that is what we must struggle for. Not now, now in the pale shadow of the future, although life today is much better than in the past, but tomorrow, a tomorrow which is not far off.

Think where we would be today, if this could have been said 40 or 60 years ago when the Republic was born. Then the Revolution would not have had to dedicate a whole year to the problem of a million illiterates. Our countryside would now be full of riches, of comfort and conveniences.

It is clear that we Cubans were not ready for this 60 years ago, and for that reason we have these problems today. Nevertheless, we can now say: we are beginning! For this reason we will have in the future what we do not have today, what the previous generations could not do for us. [*Applause*]

Is this, or is this not, the way a people ought to think? Is it, or is it not, the way everyone ought to think? [*Affirmative exclamations*] You know that these are not simple words; you know it because now, in your children, you see that which you would have liked to have when you were children. There are many of you who say: "What wouldn't I give to be your age today; the opportunities that you now have which I did not have!" [*Applause*]

When parents who could not even go to school—who walked barefoot; who perhaps even saw a little brother left to die, without a doctor, without help—when they look at their children with hope, and above all with assurance as you all do, without a soul who could hesitate or doubt it, what parent today doesn't think that his child has a secure future? What parent today doesn't know that his child will have a very different future from the one he had? [*Applause*] What parent doesn't know that his child will become all that he wishes, that he will reach the highest places in skills, in culture and in work, and that all opportunities will be open to him. Now there is no longer even the fear that so many parents have when they ask "what will become of my children if I lose my life?", because they know those worries are a thing of the past.

Thousands of young campesinos are now studying: if there are not more in our Institutes, in our Technical Schools, in our Universities, it is simply because most campesinos did not go beyond second or third grade. That is why there are not more campesinos in the secondary grades; that is why there are not more campesinos in the Universities. [*Applause*] Nevertheless, soon it will not be like this, because soon he will be able to go as far in school as he wishes.

And if there are not enough teachers . . . I can now say that the appeal made by the Revolutionary Government for 4,500 students for the Pre-Vocational School of Minas del Frio has met with a response of 8,000 applications. [*Applause*]

Although our schools might have deficiencies, teachers deficient in some case and even teachers who do not teach, it will not be this way in the future, because everything is being built from the ground up. And although there are not thousands of young people from the countryside in our Universities there are thousands of young agricultural workers studying. And this year, in the next few weeks, we will have 2,000 students in the Soviet Union [*Applause*] studying problems of administration, of machinery, of agricultural technique, and 3,500 studying problems of administration in the capital.

Therefore, these 5,500 students added to other schools of this type, which were already functioning, form a total of 6,000 young farm laborers. If we add the School of Insemination, the School of Revolutionary Instruction, and the Young Campesinas (farm girls) to this figure, this makes a contingent of more than 10,000 young campesinos. That is, 10,000 young people coming from the centers of the Granjas and from the old Cane Co-operatives. Ten thousand!

This gives you some idea of how opportunities are opening up everywhere for a young worker to become a technician, for a young worker to become a director of an agricultural enterprise; that now it is not the Yankee master, nor the exploiting landowner, but the young who have merit and capability. Because if there is something that we ought to understand, it is the necessity to form competent examples, to shape men, so that they will not fall into our errors and incur our faults.

What is it? You know! If there have been many faults, many errors, what other way is there to overcome them? No one is born with knowledge, and many men who were called upon to do a certain job that they were not able to do, were not even able to say that they did not know.

And the guilt, we ourselves cannot brush aside either, because we ourselves are guilty. [*Applause*] If within the next few years there are no perfectly capable and competent men, then we ourselves must accept the blame. But the guilt will not fall on us, because we know what we are doing. We know that in the future we will not have the difficulties, we will not lack the elementary things that we lack today.

Today, it is the bitter gift of hard work, suffering and patience, which requires all the fortitude of revolutionaries, encouraged by a tomorrow which we know will be very different [*Applause*], when the masses of young people, truly prepared, become part of the task, become part of the force.

Tomorrow there may be other problems, corresponding to new stages of progress. All that is lacking today, will be abundant tomorrow. But it is not a question of days, nor of weeks nor of months. It is a question of years!

Of course we all wish that tomorrow were here; we all wish it immediately, but nothing like that happens in real life. The seed does not bud sooner by wishing. It always takes years. Thus as the parents look at their children, at the little ones recently born, they are not impatient. They care for him, knowing that some day they will have a man in the family. Thus also with the faith that we have, that work today [*Applause*], the work of looking after the Revolution, will create tens of thousands, hundreds of thousands, yes millions of new men to the family of Cuba. [*Applause*]

Only with work and sacrifice

There are many things about which we ought to think seriously and responsibly. There are many evils, defects, vices against which we still must struggle, in order to merit a better future, which can be accomplished only by sacrifice. We will not accomplish it by sleeping in the shade; we will not accomplish it if we are like vagabonds or loafers. Abundance of all that we wish, of all that we need, can be attained only with sweat, with work and with sacrifice. [*Applause*]

For that reason, compañeros, we must take the spirit of the Revolution and of truth, to all the workers of our country. We must make them conscious of the duty of work, and that work is not a punishment, but a necessity in the life of man. That which makes a man a man, and distinguishes him from the others, [*Applause*] and makes him lord and master of nature, is work.

The vagabonds do not progress. The vagabonds will never help us to liberate ourselves from want and misery. For that reason, it is necessary to create a devotion to work; to see work as it is and not as a punishment. In the past, work was an instrument of exploitation of man. Today, it is an instrument of the redemption of man, of the elevation of man, of the progress of man. [*Applause*]

We know that there are many things to overcome, many deficiencies, many things which grieve us, errors which grieve us, weaknesses which grieve us, carelessness which grieves us, as, for example, uncultivated lands and shortages of agricultural products caused by carelessness, because those in the front line have not been attentive, have not listened, have not paid attention to plans.

With an attentive eye on all problems, we fight to conquer them the way we resolve the problems of supplies, without stepping backwards. We use as an example the problem of the supplies of those products themselves. We have thought much about this problem and have discussed what should be done. Should we give a plot of land? No! Because he who has a small plot wishes a larger plot later; his animals multiply and he no longer has three, he has 10 or 20 or 50, and the worker turns up as a latifundista, because every herdsman must have pasture for his particular herd. [*Applause*] No! We must not use those methods which encourage the abandonment of the obligations of work. We must not go to individualism which encourages egotism, which encourages inequalities among men; we must go to the collective. . . .

How do we resolve the problem of housing? It is impossible, now, to build houses for all. It is necessary then, in order to resolve this problem; not to be too ambitious, and spend whatever we can to resolve the problem of a place to live at least, even though that dwelling is not as good as the houses being constructed in the towns, because the housing problem cannot be solved in one year.

Each and every one of those problems must be looked after. But how? By thinking of the future, thinking of the interests of the Nation, thinking of the interests of all the workers. This is how we must think, because if we are all dependent on others, if no one can depend only on himself, then we must always think of the interests of all. [*Applause*] If one concerns himself not only with his own needs, but with the needs of others, then others will look out for the needs of the individual. [*Applause*]

Thus we must discuss, not with orders and commands, but with reason and truth. Because faced with the truth, faced with what is reasonable, no one can oppose; faced with what is just, no one can be opposed. Always with reason, always with what is just, always discussing, always teaching, never imposing, but persuading, with you participating.

From among you must come your union leaders. Now, at last, we can reply to the question which has been asked many times: "Why don't we have a union?" Indeed, you will have unions; your union leaders [*Applause*], the Technical Advisory Councils will come from among you. In the future, more and more of you will become leaders. And finally, from you, from the mass of the proletariat, will come those who will advance the countryside. You will display the maximum interest, responsibility, feeling of duty, patriotism; thinking of the nation, thinking of our great people, who united must march forward, who united must conquer their future. [*Applause*] Thus each time more aware of our social debts; each time less egotistic; each time more brotherly; each time more stripped of the dead weight, of the vices, of the past, to continue adapting our thinking and our actions to the present and to the future. . . .

We trust in you, cane workers; we trust in the revolutionary spirit of that great mass. We know that you will put forth the maximum effort in the face of the present difficulties. [*Applause*] We know that you will assert yourself in the face of weakness, the spirit of laziness and those who do not feel a sense of duty, those who do not understand the great truth: that work is the most honourable activity of man, the most fundamental necessity of man; that work makes men out of us. [*Applause*]

Let us work for all, so that all will work for each one of us.

Our Country or Death!
We Will Win!

SPEECH TO THIRD NATIONAL CONGRESS OF NATIONAL ASSOCIATION OF SMALL FARMERS
by Fidel Castro

[*The delegates greet the speaker by chanting: "Fidel, for sure! Hit the Yankees hard"*]

Invited Comrades,
Comrade delegates to the Third Congress of the ANAP:

The Yankees have to be hit hard on many scores. But they also have to be hit hard in agriculture! Hitting the Yankees hard in our agriculture will mean the defeat of one of the principal weapons that they have been using against our Revolution, that is, the weapon of economic blockade; in other words, the weapon of hunger.

That is why, since our Central Committee, in the statement published today, expressed its opinion concerning international political problems, we wish to speak here tonight of agricultural problems [*Applause*]. An excellent way of hitting the Yankees hard!

Just a little more than eight years have passed since the triumph of the Revolution, and we believe that after eight years both you, the farmers, and we, the leaders of the Revolution, are in a better situation than ever before to speak together and understand one another. Because the farmers know a great deal more about revolution than they knew eight years ago and the leaders of the Revolution know a little more about agriculture than we knew eight years ago.

And if you know a great deal more about politics and we know a little more about agriculture, then our understanding one another becomes a very simple matter.

I do not intend to imply that we knew much about politics at the beginning of the Revolution. And we must say that we also have had to learn a great deal during this period of eight years.

Can we say it was easy for us to understand each other's ideas at that time? No, it was not. You knew very little about politics and so did we—when I speak of politics I mean revolutionary politics. We knew nothing about agriculture and you knew a little more than we and yet you, too, knew very little about it. I trust you will not be offended by my saying this [*Shouts of No!*] And even now, we still know very little.

I know that someone may rise to his feet and say: "I am an expert in such and such a thing", and someone else may stand up and say "I am an expert at this other thing." We have some experts; there is no doubt about that. We have some very good farmers who know a good deal about tobacco and others who know a good deal about sugar cane or citrus fruit trees, while others know about cattle raising. But when we speak of knowledge we are speaking, in general, about the knowledge of the masses, of the immense majority.

I am not going to speak about such problems now. I would prefer to speak of the problems that made it difficult for us to understand each other. The farmers always had great trust in the Revolution from the very beginning. That is, the farmers never lacked faith in the Revolution. What was not clear was a problem about which relatively little experience existed anywhere in the world, a problem that, for that matter, had never been fully solved anywhere in the world: the problem of agrarian reforms.

What did an agrarian reform consist of? Even

Speech at the closing session of the third congress of National Association of Small Farmers (ANAP) in Havana, May 18, 1967. The translation is by the Cuban Department of Stenographic Transcripts.

today, all over the world, there is much talk about agrarian reform. Everyone has a different concept of agrarian reform. But the fact is that our young revolution was faced with one of the most difficult tasks in agriculture: how to carry out an agrarian reform.

Some said: "But it's very simple; distribute all the land." In fact, until that moment, the most prevalent idea in this country of how an agrarian reform should be carried out was that all the land should be distributed.

And really, today we can clearly see that that idea would have been perfectly all right for a capitalist society. For in capitalist society it would have been impossible, technically and productively, to do what can be done with the land when all the resources of the nation are utilized. This type of agrarian reform, moreover, would have been the easiest of all to begin with: so many thousands of hectares, so many farmers or farm workers without land. How to distribute it? In plots of one **caballería** (13.4 hectares). It would not go around. There were hundreds of thousands of persons living or working in rural areas who wanted land. Half a **caballería**? It wouldn't go around.

A quarter? An eighth? Of course we mean arable land; it wasn't a question of parceling out the Pico Turquino, Cayo Romano, Cayo Largo or the Zapata Swamps, but rather arable land in plots of one-eighth of a **caballería**.

And if we had distributed the land in plots of one-eighth of a **caballería** . . . We can say today, though five or six years ago we couldn't have made such a categorical statement, that if we had distributed the land in plots of one-eighth of a **caballería**, this Revolution would have gone to the dogs. And, of course, since we did not intend to go to the dogs, we would have had to about-face the process: begin to pool all the plots of one-eighth of a **caballería** again, because it would have been absolutely impossible to apply a machine, a combine, install a whole hydraulic system or fumigate or fertilize by plane when this was desirable.

And naturally today we know this. When we undertake the cultivation of the land, when we extend the area of cultivated land, when we increase the number of milk-giving cows, when we increase the amount of sugar cane cut. when we increase the quantity of coffee, there is a shortage of labor, and the need for machinery arises. Furthermore, we could not have used machinery and the application of technical methods would have followed an opposite course if we had begun by distributing the land.

What solution did our Revolution seek? The farmer, who has been working the land for years, who is accustomed to this way of working and is paying rent, paying one third, or paying 50 per cent—all these forms of exploitation which existed for farmers—this farmer is used to that way of working, this farmer has his whole mentality adapted to that way of doing things. He even prefers it. We left that farmer there, we freed that farmer from his rent, that is, from exploitation and we began to give him all services possible—education, doctors, credit, communications—in fact everything within our power to give.

But why should an agricultural worker, why should a laborer, who one day can be a worker on the land aided by major machinery and technical methods, become instead a minifundist (marginal farm owner). Because we must make a distinction. There is the farmer who has 10 or 20 hectares and who can develop his crops. There are even an infinite number of workers who have a small holding here or there and whose crops are actually for self-supply. This is why the revolution decided not to divide up the big latifundia. It was not an easy task; it was not an easy thing for people to understand. The bourgeoisie said, "Look, the State is taking over the land; look, now you'll be wage laborers for the State."

After that, since there were peasants with different amounts of land, whenever an agrarian reform was carried out, they tried to stir up people's anxiety. Naturally, it was absolutely necessary to carry out a second reform. Why was it so vital? Because a large part of those landowners who still had between 250 and 500 hectares were virtually sabotaging production. It was necessary to carry out the other agrarian reform.

What did they do then? They began to say, "Your turn is coming next." Then the Revolution stated—and this Revolution has been characterized by doing what it says; it has been characterized by its seriousness, for keeping its word—, the Revolution said "there will be no more agrarian reforms." [*Applause*]. The process of agrarian laws and reforms lasted up to that moment, in fact.

The bourgeoisie in general and the landowners used many arguments. They said to the peasant farmer, "This is socialism, and that means they're going to socialize your land." We came along and said to the peasant farmers, "This is socialism and that means we're not going to socialize your land." [Applause] Because socialism is a realistic and scientific conception of society, and because the poor and exploited peasant is definitely an ally of the working class. And since the poor and exploited peasantry is an ally of the working class, that peasantry must be treated as revolutionary, it must be treated as a comrade, as a friend, it must receive all the political considerations to which it is entitled [Applause].

Because, what is a revolution? The social revolution is the close union of all of the exploited against the exploiters. And a peasant who worked his land and paid rent and had so many other burdens was among the exploited; that peasant didn't exploit anyone, he worked the land with his own hands and was exploited; he necessarily had to be an ally of the revolution. And in our revolution especially, the peasantry played a very important role, as the first guerrilla nuclei were formed in the mountains, among the peasants [Applause].

I believe that by this time not one single gullible farmer can be found

The reactionaries' arguments, trying to stir up fear and confusion, might have influenced some, but I think that by now the farmers cannot be deceived by any of those tales, and for that reason we can speak clearly, with perfect frankness [Applause].

We feel that the Revolution's solution was a good one, both in its decision not to divide up the large landholdings and in its decision to respect and maintain the poor farmers' traditional means of production. And we have never made any attempt to establish socialist production among the small farmers [Applause].

We especially recommended not fostering cooperatives. Why? Because you begin to form cooperatives and those rumors gain force—and I refer not to the credit and services coops that you all know and have organized, but to the procedure of uniting parcels of land—if that is done, the campaign will prosper, the lie that we want to socialize the farmers' land.

Of course, it is true that a large parcel of land means much more production and greater productivity. But we decided that didn't matter. On the large landholdings there is so much territory covered with brush and briar fields—producing practically nothing in that state—as to provide for an extraordinary increase in production with cultivation, although production cannot be increased to the same extent among the small farmers because of the land division.

Today we see clearly that this was the best policy that could have been established.

Does that mean, then, that we believe in minifundia, No. We don't believe in small landholdings. Does that mean, then, that we think maximum yields and maximum productivity can be attained through the cultivation of small parcels? No, it doesn't mean that. It means that the Revolution is following a truly realistic policy, that the Revolution is following a correct policy, based on the realities that exist in a definite country, based on the situation that exists in a specific country. One thing frequently occurs to me. Every time I travel through the mountains I suffer terribly. I suffer terribly because I've seen the colossal destruction that man has wrought in the mountains. Every time I travel through the Sierra Maestra, the Escambray mountains, the mountains of the Second Front, many of the regions of the country, I can't help feeling sad to see how man has been destroying nature.

And that nature is the nature that other generations will have to live off in 20, 50, or a hundred years; it's the nature that double, triple, four, five, ten times more population will have to live off in the future. We ask ourselves whether this generation of Cubans has the right to destroy nature? Do they have the right to leave future generations barren rock? And naturally we have to admit that we don't have that right.

But we also ask ourselves whether that Cuban, that farmer, is to blame for having been forced to commit that crime against nature? No, no! What was it that forced that human being, that man to climb to the top of the hill to cut down the forest, to burn the timber, to plant anything one or two years, while the rains came and carried off the topsoil? What forced him? Did he go there for the fun of it? Did he go there aware of what he was doing? He was forced to go there by an inhuman social regime, an exploiting social regime, a selfish regime. If that man knocked on a door asking for work, he didn't get it. If he knocked on a door asking for bread for himself or his children, no one gave him bread, no one gave him work.

That society condemned man to live as he could, to die of hunger if he could not live. And as the population grew, that forced masses of men to seek refuge in the mountainous zone and to begin there, with no resources whatsoever, with no roads, with no credit, with nothing! They were forced to go there and plant some crop in order to live. Naturally, population growth continued during that time.

New cement factories will solve the housing problem

Moving through the Sierra during the war, we would come to some truly inaccessible places and see a man perched up there at an 800-meter height, working there at a 70 or 80 degree angle. When we saw that man working so hard, we would say to ourselves: "How many times have our farmers been slandered, called shiftless, lazy, and just look at that man, almost among the clouds, almost tied there in order to plant something!" And we also said to ourselves: "What will happen in this country when the few remaining hills where man has found refuge are all occupied, are all barren, all eroded, for then there will be nowhere but the sea where refuge can be found?"

And when the Revolution triumphed there was practically no place where the land was not occupied. Those men were forced by that social regime to destroy a good part of the natural resources. Obviously, the forests had disappeared.

We see, for example, that housing is one of the most serious problems in our rural areas. It is a serious, difficult problem that we'll solve because we are setting up new cement factories. But lumber to make a door, to make a table, or even a coffin, is not available. And how could it be available? To explain the lack of lumber in this country it is necessary to go to the mountains—not to the plains where there haven't been any trees for a long time. How the mountains were stripped, how the woods were stripped! For there where the land wasn't good for farming—for example, pine forest—and peasants hadn't settled, in went the land-grabbers, the timber barons, who chopped down hundreds of millions of trees, without planting a single new one. There are surely farmers from the Sierra Maestra present here, farmers from the Alcarraza zone, from Pico Verde, Pino del Agua, Pinalito, they surely know what lands I'm talking about. Those lands are generally lateric, red soils, where only pine trees grow.

We have already learned to make even those lands produce, as is the case with Pinares de Mayarí. But of course those lands were stripped of the forests and no one worked them.

Nevertheless—and this is incredible—we have at times seen farmers who were such optimists that they went to those hills—those pine lands—to plant root vegetables and plantains. But of course the plantain there grows to a height of one or two meters without producing a single fruit; no root vegetables grow there either. The most sensible thing, really, the only sensible thing to do with those lands is to plant forests. And that is just what we are doing.

I told them that it's not that we believe in the **minifundia** as the ideal method of production; furthermore, that society forced illegal use of the land, often forced criminal use of the land. It's not that our other land is always correctly used. What happens is that the effect of erosion on level land is much slower and naturally, not the same as on steep mountainsides where topsoil disappears completely in a very short time. Much of our level land is affected by a certain amount of erosion, but in general its effects there are not the same.

We believe that the small parcel of land, both from the standpoint of the land and from the standpoint of labor, is not the most rational, the most productive form. Nevertheless, our policy was to maintain those parcels, our policy was to be patient, to struggle to introduce technology—although it has still not been sufficiently introduced—into that mode of production, respecting the farmer's desire to produce in the way to which he was accustomed, in the way he chose.

The farmer has his own mentality; very often he doesn't even want to have a neighbor too close; the idea of a house being built next to his own horrifies him. Many farmers think that way. They don't want problems or arguments with neighbors, or with the neighbors' families. They are very happy living alone, isolated. Of course, it is not the ideal way to live. And often when one passes by those isolated places he observes that the ones who suffer most from this are his wife and children.

They are the ones who suffer most from the isolation in which farm families generally live.

Those who suffer the most from the life of isolation often preferred by the farmer are his wife and children

One day during a tour we went with the President and other comrades to a place called Purial, El Purio—between Mayarí and Moa. We saw some steep mountains and some pine woods; we were working on some forestry plant there, so we went down a forest trail and an hour and a half later, in the most remote, isolated spot, we came across a crude hut.

And out of that hut, just a meter and a half high, emerged a man, and then a woman, and then one child after another [*Laughter*]. It was like one of those circus acts you've all seen where half a dozen people emerge from a tiny automobile. And that's what happened there. There was a doctor with us, who began looking at the yellowish, sickly children. The parents began explaining the problem of the children, who turned out to have parasites and worms. The doctor of course saw right away that it was a question of parasitism . . . The place where they live is very remote—no vehicle ever passed that way—and those children hadn't been taken to the hospital, even though there was one just 15 kilometers away. He took note of this immediately and later had a jeep sent to fetch the children and take them to the hospital. But the doctor said, "Well, within a few weeks the children will be in the same condition." Those children were astonished to see a motor vehicle. Imagine! Right there in the middle of that woods, isolated from everything, lived this charcoal maker.

Of course, not every situation is that extreme, but we always notice the children and what their problems are: they have no place to play, they live in such isolation. And we see the women's problems: the washing, the cooking, water—often they have to carry it a long distance—especially in the mountains, it's really a tremendous problem.

This isolated life is often preferred by the farmer. True, those who suffer the most are the farmer's wife and children. The situation isn't ideal, but nonetheless, there is always a solution; there are schools, there are scholastic programs, there are medical services, there are communications, and there are boarding schools in the mountains. In short, even in this situation there are a number of programs that can considerably improve family circumstances, and the circumstances of the children.

In the long run our policy has been and will continue to be one of complete respect—complete respect for the wishes of the farmer to work the way he sees fit, so long as he sees fit.

We wonder if there will be small farmers in 40 years. And the answer is that if in 40 years farmers still exist who want to be alone, isolated, working there with a yoke of oxen, with a very low productivity, who prefer to stay that way, we'll leave them there even if it's 40, 50, 100 years [*Applause*].

Does that mean then that this will last forever? No. It won't last forever, but that won't be because of any law of any kind. It won't last forever because of the incredible, colossal development in the agriculture of this country, in Cuban society, because of the tremendous development of technology, because of the fantastic development of social and educational programs. We've already seen innumerable cases of farmers who lived in the mountains and . . . with two sons in the army, two daughters here studying under a scholarship, another a nurse, still another a teacher there, and they've been left alone and they say: "Look, I'm very old, everyone's left me and the fact is that we'd like to move, we'd like to sell out." In the Sierra Maestra, since we didn't want illegal deals to be made, we decided to authorize and facilitate resources to permit the ANAP to legalize even some illegal cases which already existed—because there were some who'd left, others that had squatted, we'd again accumulated a pile of illegal cases. We're going to legalize these cases—some sales that had been transacted—once and for all. Then those that want to leave, well, we'll buy them out, because if we don't, we're never going to be finished in these mountains. The same problem would go on forever. One leaves and another arrives, and every time this happens there's less vegetation; those mountains are stripped more and more: Those who want to sell, we're going to buy them out. Of course, who would have sold 10 years ago!

But now, naturally, many of those farmers have sons studying technology and they have many opportunities down below, although not sufficient facilities. And since sales were authorized, in a matter of a few weeks four thousand farmers sold and we had to ask the comrades in the ANAP: "Look, hold back a bit. Hold back a bit because they are leaving the Sierra abandoned." Then it was a different problem: shortly the land was going to be uninhabited, and not only that.

We expect some day to have all mountains reforested

We said: "Buy from those who are alone, from those who are already old, from those who cannot work. If a farmer is young and can work and wants to sell anyway, tell him, no, to stay up there, that he is needed up there, that the coffee must be tended; that it is important." Some sold and later set up a fried food stand, or a fritters stand, in a wagon. Gentlemen, that is going backwards, that is going backwards! We are better off if that farmer is there, even though he is producing little, on the side of the hill . . . We really prefer to have him on the side of the hill, even though it deteriorates the hill, than to have him come down off the mountain, to become a fritters vendor, so that, instead of the hill, the farmer deteriorates. Because he who stops producing with his work to become a fritters vendor, ceases being a worker to become a businessman; because it is easy, in any place here, with the money that the people have . . .

It's the same with the taxis: some of these jeep jitneys around here—who knows better than you?

How much do they charge? How much do they charge? Whoever has a jeep—because unfortunately there is still not enough transportation—whoever has a jeep gets rich. Then an individual with a little car, or with a jeep, earns 40 or 50 pesos. He can earn in one year, three times as much . . . did I say three times as much . . . no, ten times as much as a farmer cultivating one half a **caballería** of cane and working very hard to cultivate the half **caballería** of cane. That really is privilege.

Anyone who sets up a fried food stand, or a fritters stand, things they can make with black-eyed peas [*Laughter*] and a little lard bought on the black market. Or rather I get lard from a little pig I've bought, set up a fritters stand, and earn 50 pesos. Naturally, if everybody set about selling fritters, nobody would get even 50 cents. And for us it is going backward for the farmer to come down off the mountain and to set about selling fritters; it is going backward.

There is another problem: the housing problem has not been solved. We don't have the means yet to say: "Look, now if you want it, here is a house to live in, a pension; we can give you all that." It is for this reason that we tell the comrades of ANAP: "Wait, wait, this can't be done suddenly, in any old way. In the case of families who, because they are old, because they want to retire, because they have children studying, want to live in the city, or go to a State farm, or have a house in a place where they can live out their retirement, nearer their children, it must be done in an orderly fashion, it must be done when it can be done. Comrade Pepe insisted: "There are still many elderly peasants in the mountains; we need to be authorized to solve their problems." I said to him: "All right, we'll make a thorough study of the matter; make a list of all the people who are in that situation . . ."

What does this mean? That none of these historical problems such as emigration from the mountains, the denuding of the mountains, can be solved from one day to the next. We have reforestation plans. We expect one day to have all the mountains once again covered with forests.

Will coffee trees be lacking in the mountains? No, not at all; coffee production will be a by-product of the mountain forests. The woods will be the main thing. We want to see the day that we can send to the farmer there in the mountains—and that day is not far off—all that he needs, even root vegetables. We want to be able to say to him: "Don't go sowing plantains on that hillside, we'll send you plantains every day; the stores will be stocked with plantains, rice and beans, as much as you need. You are a coffee grower and also a producer of hard woods, you tend your trees."

In a word, over a long period, we will reforest our mountains.

The most needy cases—I was saying to comrade Pepe Ramirez—will have to be studied, but those people shouldn't come down from the mountains if they don't have a place to live below; that doesn't solve anything. That's no way to solve any problem. That's why we proposed that all the cases of peasants who want to leave the mountains should be studied.

What does this mean? That the very development of the Revolution, the new living conditions that are being created, will progressively turn those mountain lands into forests, and the lands will be gradually transferred to the National Land Fund.

This explains why, when a farmer wishes to sell his land, the Revolutionary Government requests option to buy it. In other words, we do not want the number of small farmers to increase. Let those who want to sell, sell to the nation. Let that land become national land.

What does this mean? It means that within 30, 40, 50, or 100 years—sooner or latter—the day will come when the ANAP will have disappeared. Does this worry you? [*Shouts of "No"*]. Who among you expects to live over a hundred years? [*Laughter*].

But you all understand well what this policy consists of. Later on we are going to speak a bit about agriculture, but now we are speaking of the Revolution's policy regarding agriculture and agrarian reform.

How can we supply an increasing population unless we use machinery?
That is, that some day, as a result of this process of evolution—and we expect this to be accomplished, as one of the most earnest promises of the Revolution, without any new agrarian reforms, ever—the farmers' children will become technicians, acquire new habits, a different mentality, other concepts. It is already happening: we have had to send many young people into agriculture while many of the farmer's children are attending school. And the son of that farmer who spoke here tonight will not live in a minifundium—not because he scorns that kind of life, no; not because it would be unbearable for him to go back to a hut with a dirt floor, no; but because he understands that he cannot produce food for the people with a team of oxen, because with that team of oxen all he would produce is food for himself, his family and perhaps a few others.

How can we supply an increasing population unless we utilize machinery? Yes, of course, we are already introducing such machines as the tractor. Things get more complicated when we start using the airplane. We are already spraying sugar cane with foliar urea. This is a superb procedure. A plane needs only one day to spray 1,340 hectares of land with foliar urea. Do you know how many men are needed to apply the same amount of urea by hand? At least 2,000 men, 2,000 men! And we need 50 trucks to transport these 2,000 men, 20 fertilizer deposits in 20 different places, and the fertilizer to be transported to all of the fields must be loaded into sacks. In case it has rained and the fertilizer is to be applied immediately, since nitrogen must be applied to wet ground, what happens then? Let's suppose it rained this afternoon; in three days that moisture will have been lost and the land will be dry because it is impossible to mobilize tens of thousands of men to apply the nitrogen immediately after the rain.

By using a plane, one man can apply the fertilizer to 1,340 hectares in just one day and at the right time, when the soil is wet, and the only aid he needs is that of the ground crew at the airport—the only place where the nitrogen (urea, in this case) is stocked—plus the transportation of the fertilizer, via truck, to the airport. Last year, when application of foliar urea was begun, there was a pilot who sprayed over 2,000 hectares of land in one day. Three thousand men could not do that job on the ground! [*Applause*]

Naturally, when a farmer's son enrolls in a technological institute and learns all the modern techniques, he draws his conclusions and says: "Imagine, when even the oxen have been or are being liberated from hard work, why should a man go on working like an ox?" What does a man produce by working like an ox? He produces a little more than an ox, just enough for himself and his family, and it's grueling work.

Those sugarcane farmers, for example, who have now had the opportunity to get to know the cane-loading machine, no doubt remember the time when they had to rise at 2 a.m., yoke the oxen, go to the canefield, load the cane, bring it to the loading station, go back to the field and cut cane all day in order to keep feeding the loading station. The farmer was forced to work 15, 16, or 17 hours a day. Did he ever get rich? No. He never got rich. And he had to do a type of work that the human body cannot stand. Well, did the farmer die? No, he didn't die, but how long did he live? It is not only a matter of asking if he died but rather how long he lived; what the life span was of a man subjected to such brutal, exhausting work.

In the past, during the sugarcane harvest, the canecutters had to work 15 or 16 hours a day, because once they cut the cane they had to load it, stalk by stalk, into an ox-cart. Can a nation become rich when people have to cut cane by hand and load it into an ox-cart stalk by stalk, working 15 or 16 hours a day? In addition, there were so many unemployed that the men were not even allowed to work the full 15 hours. There was not enough work for so many people, so one hand was forced to work fewer hours. But when given the opportunity, he would work the full 16 or 17 hours. Agricultural workers in this country used to load forty million tons of sugar cane, one stalk at a time, into ox-carts! Today, more than half of our total cane crop is loaded mechanically and this year the small farmers too have new loading machines for the next harvest [*Applause*].

The day will come when every farmer's son will be a technician

That is, this is going to ease working conditions for the men: the cane will be loaded by machines. The day will come when machines will cut all the cane, too. We still have to solve that problem; we have no other choice but to solve it; we will have to solve it, and we will solve it. But right now all the cane in the provinces of Las Villas, Camagüey and Oriente is going to be fertilized by plane with foliar urea, three times. The farmers in the northern zone of Oriente may have been able to see the results of the urea on the cane, how all the estimates of the cane turned out to be low, how green the cane became, how it grew. Nevertheless, when the situation was brought up—"Well, now we're going to begin to spread the urea."—and we planned to fertilize the cane of the small farmers, too, by plane, a problem arose—"Impossible, impossible, because a plane will have to turn into a grasshopper jumping about from place to place." What are we going to do? Well, we are going to give them ammonium nitrate. We have calculated that a ton of urea sprayed from a plane means an increase in production similar to four tons of ammonium nitrate applied on the ground. That is, applying it by hand we have to use four times as much fertilizer; and of course if a plane sprays 670 hectares a day, as an average it will be doing the work of a thousand workers a day. And right there we are faced with a problem: there is a technique that could save an extraordinary amount of work and it can't be applied. We have to go to the man, to the small farmer—working like an ox—spreading nitrogen all along the furrow, and we can't take advantage of such a mighty machine as a plane to do the same work with only a fraction of the effort that the farmer is putting forth.

I am explaining all this to you so that you will understand why the son of a farmer who leaves home to study in a technological institute, and has the chance to continue his studies in the university when he graduates, says: "No, I'm not going to do the work of an ox." He doesn't return to the minifundium. He will undoubtedly not return, and his attitude is correct.

What does this mean? That society is advancing, that the time will come when every farmer's son, without exception, will be a technician. The best proof of this: the technological institutes will take in 40,000 students this year; in 1970—just see how the plan has been growing—in 1970 there will be one hundred thousand students in the technological institutes! [*Applause*] Ten more years and there won't be a single youth in the rural areas of this country who won't have a junior high school education and have become a technician.

Do you see that this is the path? Do you see why we have to wait 10, 20, or 30 years? Do you see what a Revolution achieves? You have to see today's reality, and what tomorrow's reality will be; you are today's reality. Today's farmers are today's reality, and tomorrow's reality will be your children; tomorrow's reality will be technicians such as those about to graduate from our technological institutes and go on to the university.

They will be richer than you, of course, because they will produce 8, 10, 15, 20 times more than you, with only a tenth of the effort you put forth. They will use airplanes on a wide scale, machines, the most modern technology; the rural areas will be completely electrified. Instead of ox-power it'll be horse-power, the horse-power of the tractors, of electric motors. [*Applause*]

If all the water that the pump of a deep well draws for irrigation had to be carried bucket by bucket, what a fine fix we'd be in! Would we produce everything we needed? Would there be enough food for all our people? Today a motor runs on fuel; it could run on electricity. But it'll be necessary to have all these machines, all modern methods, all the electricity, all the energy, everything, producing for man.

That day is distant, but not so distant. The day will come when there won't be a single miserable hut in our country. Today there are still many, a great many miserable huts in our country. And just as when we see the eroded mountains, we suffer when we go through the rural areas and we see so many tumble-down huts. Sometimes there isn't even enough thatching. There isn't enough thatching because of the needs of poultry farms, dairy farms, one thing or another, endless reasons.

The coffee nurseries . . . As you know, the coffee plan is gigantic, and the amount of palm thatch that has been brought to the coffee nurseries is huge. There is not sufficient royal palm thatch. If too much thatch is cut off the palms, they don't produce nuts. There isn't enough thatch of other palms to go around, nor has there ever been. And logically, there are many houses in very bad condition.

Some day all of our rural areas will be electrified. Some day all our rural areas will be full of towns with running water, electricity, gas stoves; and the children will not have to walk even two kilometers. They will go to school starting in the morning, and they'll have breakfast, lunch and dinner there; they will spend the day in school and will return home at night. There will be no more washtub; no more carrying water [*Applause*]; no more candles and lanterns; children will be a thousand times better off; women will be incomparably better off.

Today's construction effort has to be directed toward creating better living conditions

But, just who is going to live like that? You? No, your children. It will be your children, because they are the ones who will be adjusted to that way of life; they will understand all the advantages of a different mode of production.

And what will become of today's farmers? Today's farmers will continue to live as they are now; not as badly, of course. There will be resources to improve their living conditions; they will have better communications; there will be many more schools; there will be many other things.

I have depicted two epochs for you: the past and the present. I have tried to give you an idea of what, in our opinion, the countryside of the future will be, within 15 or 20 years, when new generations will already be applying technology, working and producing in a different way from today.

Of course, there are many farmers who have a tiny plot of land; they are semiproletarians. When the new towns are built on the State farms, many of those farmers go to live in the towns immediately. And if the living conditions are still bad in the farmers' huts, they are still worse on the cane plantations. Today there are still tens of thousands of workers who live in barracks with their families, tens of thousands of workers living in one room of a barrack. Building efforts must logically be aimed at creating better living conditions for those workers.

It is true that farmers still have very bad housing, but the farmer has had many problems solved that are still affecting the worker on the cane plantations; and those workers are making a considerable contribution to the economy with their labor. It is logical—and I think you realize that it is very fair—that the bulk of materials be invested at present in improving the living conditions of the workers on the State farms [*Applause*].

Now you will say: Well, those are problems of the future. And the problems of the present? The present picture? What is expected of us? How are we going to produce? How can we make our efforts more useful, the best possible? How can we make our lands serve both our families and country to the maximum extent? These issues of today are, of course, the ones that are of most interest: what we want, what we expect of the small farmers, how we view the process of development of our agriculture.

A few years ago the State farms were a disgrace because too many inexperienced people—as we have explained on other occasions—often with little more than good will to recommend them, kept production on these farms at a poor and deficient level. These farms had, of course, had the advantage of maximum resources, not in order to grant them privileges over and above the farmers, but because they included the greatest extensions of unproductive land, of brush and briar; and besides, speaking frankly, the production of foodstuffs on these farms was much more assured than in the case of the small farmers.

Let me explain. There are all types of farmers. There are those who are rigorously honest, who work with tenacity, who don't aspire to personal enrichment, through selling things three or four times their value; and there are farmers—above all in the areas nearest the cities—who, if they produced an egg, sold it for 30 cents; or if they raise a hen, today, they sell it for 5 or 10 pesos.

And remember that the policy of the Revolution has never been to prohibit farmers from doing this. Any farmer may, if he chooses, stand by the roadside and sell his produce, whether turkeys or chickens or anything else. No one bothers him. And how much does he charge for these things? Well, sometimes five times their value.

Many people in and about the cities drive into the countryside and buy three, four, five, ten quarts of milk—since there is no regulation on this—paying four times what it is worth. And you all know of those "operators" who pay a peso for a quart of milk, since that peso doesn't cost them anything—it has been obtained through sheer robbery, to speak frankly. [*Applause*] For that matter, anyone who sells anything—lollipops!—can easily earn fifty pesos. Fifty pesos! And such people pay a peso for a quart of milk. And in the same way, they buy everything they can get their hands on. Not all farmers have a social concept. There are some who speculate with no pangs of conscience.

The day will come when fruits, vegetables and even milk will be distributed to the people free

What is our policy toward speculators? Tranquility. Will we at some time arrest them for engaging in this form of robbery? No. That is not our way. Let the man who chooses to stand by the roadside and sell a turkey for a hundred pesos do so, and let the farmer who wants to sell milk at a peso a quart do so. It doesn't matter, let him sell his milk at a peso a quart!

I am going to explain the way that this problem will be solved. It is simple, very easy. The day will arrive when this individual will run out to tell the driver when the milk truck passes, "Listen, please collect all my milk." And do you know why? I will tell you why. Because the day will arrive when fruit, vegetables, and even milk will be distributed without charge to the public. [*Applause*]

We know what we are doing, and we know what the production levels will be in this country within a few years. We know how many cows are being inseminated. We know how many calves are being born and we know how much milk a first-generation cow from a Holstein and a Zebu will give, and we can calculate how much milk we are going to produce, how much fruit. We know how many coffee bushes we are planting. The time will come, gentlemen, the day will arrive when we will be able to say to the people, "Go and get the coffee you need at the market, without charge." [*Applause*] Why? Gentlemen, the coffee-growing program we have outlined is so extensive that it is enough to state that this year over 200 million bushes will be planted, and that by 1970 we will have planted a thousand million coffee bushes. [*Applause*] We know how much coffee, how much milk, how much of everything we shall have.

And when we are producing millions of hundredweights of coffee, no one will need to get into a truck and, with all kinds of difficulties, go up to where he can buy a pound of coffee and return in order to speculate with it.

Today, the farmer who works his farm is supplied with fertilizers and given credit; roads are being built for his use; free medical attention is provided, all without cost to him. And many times it is necessary to plead with him: "Listen, give up at just one sack of coffee, because the workers want to drink coffee, too."

Sometimes we meet certain farmers in the mountains and they say: "What about shoes? Those shoes are worn out." I say: "Ah, so your shoes are old. But you drink coffee 7 times a day, and the worker who makes the shoes, poor fellow, would appreciate it if you sent him a little more coffee." Each one has something to offer, gentlemen. [*Applause*]

I simply wish to explain to the farmers all these very clear and obvious facts. We would be hypocritical and we would not understand each other if we came here just to talk about the good things, the merits, the patriotism of our farmers, without stating at least some of the defects of certain farmers, of some farmers! [*Applause*]

Very well, the day will come when on each one of the many, many roads that are being constructed now . . . And there are a number of farmers here from Victorino, or from Matías or from Bernardo in the Palenque zone, or from Bayate in the Guantánamo zone, who know at what a pace the road-building brigades are constructing roads which are really highways. By the month of August, we expect to have 40 brigades building roads in the mountains and in the countryside. Forty brigades! Our countryside will have many roads. And trucks will travel these roads and then the farmer will put up a sign saying: "Please stop and pick up some of the coffee I have on hand." Why? Because there will be a surplus of coffee.

It is enough to state that we are planting 100,000 hectares of irrigated citrus groves—these are already in the process of being planted—and on these 100,000 hectares coffee bushes are being planted between the fruit trees. And it is enough to point out as well that, in the extensive reforestation program, in many areas where the soil is good, we shall plant coffee bushes under the trees. When the number of coffee bushes must be reduced in the citrus groves because of the growth of the fruit trees, the coffee bushes in the forest will begin to produce and no one will need to worry about the nuisance of papers and contracts. [*Applause*]

The elimination of those contracts seems to me an excellent idea. The farmer is not going to acquire an education from a contract, gentlemen. The people are not going to learn their duties and obligations by being obliged to become legal experts. "You signed this. Deliver one liter more of milk; deliver one liter less."

We have known how to be patient and we believe this is all to the good. It gives us moral authority; it gives us the right to be able to speak with complete frankness, with complete sincerity. We know what is needed to make all these problems of speculation disappear. Eggs are a good example. And what happened in the case of eggs will also happen one day with everything else, including chickens.

We will be producing as many chickens as the country can consume. We are not doing so now because we have had to concentrate on egg production first, to permit a wider and better distribution of eggs. But now we can step up production.

The only way to train a man to be superior is to teach him to work from a very early age

With all the surplus products for export that this country will have in future years, we will be producing chickens for consumption in the same way that we are producing eggs now—in astronomical quantities. Then the sale of chickens along the highway will disappear. The sale of chickens at the roadside will be a thing of the past! Because when a citizen receives the products he needs—and he will receive without charge all those products that we shall have in enormous quantities—the roadside stands will be finished . . . The roadside stands will come to an end, because no one will buy from them when the products are available without charge.

And already this year, when we had surplus of cabbages, these were given away free. If citrus fruit is in surplus supply at the end of the year—since we are fertilizing 4 million citrus trees—then the surplus citrus fruits will be given away free. And the distribution policy that will be followed is that every time there is a surplus of produce, this will be distributed to the public gratis. [*Applause*]

We are socialists; we are not capitalists. And this is where the great difference begins to be seen. Under capitalism, when something was in surplus because the people hadn't enough money to buy it, this surplus was thrown away. And while they were throwing out food, the people were dying of hunger.

The way to communism is exactly this: education is free—a reality for us today—and free medical attention—which also exists today—and free housing—the great majority of our people no longer pays rent—and everything little by little, with time and at the pace made possible by production increases through use of technology and the work of the entire people, everything will be free. Some may ask the question: "How is all that coffee between the citrus fruit trees going to get picked?" Through the "school to the countryside program," by the thousands of secondary schools that we will have throughout the nation where the students themselves will combine work and study. For there is no better way to train a man to be superior than to teach him to work from an early age.

Someone has said: "Can't we invent a machine to pick coffee?" And we have answered: "But that would be harmful to education. We can't mechanize everything. If we mechanize everything, how will we be able to teach our young people the meaning of labor?" We shall have to leave many things unmechanized; but of course not the most difficult. What must be mechanized is canecutting, road-building and much other ardous labor. But such work as the picking of oranges and coffee, etc.,—especially the coffee that is planted in citrus groves on land that is completely level—can easily be done by our young people.

So, how much will it cost to raise this coffee? Part of the cost will be in fertilizing it. We export part of the coffee and import fertilizers . . . Irrigation of the citrus groves . . . certain fuel. With what we export we can import fuel. And what else? Nothing else. Picking the coffee costs nothing, as it is a part of the education of our youngsters to avoid their becoming . . . Imagine! If this were not so, the children of the proletariat would become aristocrats without knowing a thing about coffee plants.

And really, although man aspires to raise his level of production, to apply machinery, we must also take care that he does not lose contact with nature. Thus, all of the youth of our country—all!—will receive an education that combines work with study.

It is for this reason that we can have all of these types of cultivation. The capitalists cannot do this because it would be too expensive and they have to worry about markets, while we have no market problems. How much coffee will we produce? We say: all that our people can consume, free of charge. Perhaps the Ministry of Public Health will protest: "We who are responsible for the health of the people consider that too much coffee is being consumed; people are too nervous; the population suffers from insomnia; the health standard, average life expectancy, etc." Perhaps they will even tell us that, there are many marital problems because people's nerves are frayed from drinking too much coffee. [*Laughter and applause*].

Then they will tell us: "No, no! We agree with your distributing milk free of charge, but coffee, no." And we shall reply: "Fine! Why don't you carry out a great campaign advising the people not to drink so much coffee? Let us organize the thing properly, because we are certainly not going to return to coffee rationing." In the future we must free ourselves from money, rationing and all of that.

Money is a vile intermediary between man and the product of his labor

The day will come, gentlemen, when, as a result of the increase in production, money will have no worth. It is true that from infancy we have been taught to revere money.

Any place you go, in any little town, there is always a little kid who asks for a nickel. There is no child who doesn't want money. Money buys everything. [*Someone shouts something from the crowd*]

A farmer says he hasn't been able to spend a cent for eight days. Everyone will live this way in the future. Yes, yes, just this way! [*Applause*].

From a very early age we have been taught to revere this thing called money that is an intermediary between man and what man produces. Man works. Here he produces potatoes: he must be given money in order to procure milk, coffee, sugar, clothes, shoes, etc. There he produces coffee: he must be given money in order to procure potatoes, milk, etc.,—everything except coffee. There he produces clothes, and he must be given money in order to procure potatoes, milk, coffee, and so on.

Money is an intermediary. It deserves a worse name: a vile intermediary between man and man's products. The day will come when he who produces potatoes and turns over his potatoes will receive no money, but will go to get his coffee, rice, sugar, clothes, shoes, everything he needs. We will do away with the vile intermediary of money. And that, gentlemen, is communism.

As I explain these things to you, to the small farmers, there may be someone who's saying, "Well, even if I'm 80, I'm still young." Perhaps; a better life at 80 means... There are many out there who are still strong at 80. "And I'm a small farmer: what do I do? Can I be a small farmer under communism?" We would tell him yes, if he wants to. But we would ask him this question: How do you want to live? "Well, I want to be paid for what I produce." And then we must ask him "What are you going to do with the money? It would be necessary to put up in all stores a list of the persons who couldn't receive things gratis because they received money! That person couldn't trade in the market where everything was distributed. 'Mr. So and So, from such and such a neighborhood, is exempt from all rights here, and must pay for everything.'" We tell him "Look, maybe it would be better if you spent the entire day resting." "No." "Why not?" "Because I've worked all my life! I get up at five in the morning; I can't live without working." This is what many farmers say: "I can't live sitting down." "Look, you have to go to the doctor, you have to be in the hospital for three days." "I can't be in the hospital for three days, because all my life I've gotten up at such and such a time and I'll die if I'm in the hospital for three days." "Well, then, you can't be sitting down?" "No." "Well, continue to work, turn all your products in free, and when you go to the store you won't have to pay anything."

What am I trying to say to you? The day will come when man will work out of pure habit. Understand that we are not going to raise children with the mentality of millionaires, sloths or parasites. It's true that we are dealing with a basic issue between you, the small private farmers, and ourselves. In the middle of a socialist revolution we're already talking about communism.

I want you to realize that this topic, communism, is very complex. But, complex because it's fearsome? No; most people are not frightened at all. It's complex because many people are doubtful about communism and wonder how it can ever come about.

We have no doubts. On a recent tour we made through the countryside we observed our young people. And what news is there of Oriente, Camagüey, Guane, Las Villas, etc.? What was said about Guane? The technological school girls are working there, about 15 or 16 hours a day and are enormously happy about it. We think this is impressive, admirable.

To give you an idea of how the 2,000 girl students from the Guane Technological School worked, I have to say simply that each one of you now has in Guane five citrus seedlings of this size [*Makes a gesture*], each and every one of you. You will ask us, "But, are you going to send them to us to transplant in our backyards?" No, but you have them there in Guane. We will send you the oranges, or the soft drinks, or whatever. Simply, each one of you has five plants there. It represents work, since it amounts to 40 million plantings, the total work done.

A new generation is arising that views work from a different point of view

We always used to say, "Well, it's true these 'compañeras' from proletarian homes normally have a nature attitude." Fine. But when we got to Banao, we found out that the university coeds had exactly the same attitude; when we got to Camagüey the comrades from the Party and agriculture told us, "What the students are doing is remarkable." The military cadets also were doing an excellent job, as were the technological students in general. And a little further on, we saw the "School to the Countryside Plan" students. The high school youngsters have left quite an impression on Party and agricultural leaders for their attitude toward the work—these healthy young students. The high spirits and enthusiasm they have shown in their work is incredible. And nobody was paying them. They weren't receiving a single cent.

A generation has arisen that views work from a different point of view. And those who plant fruit trees there, who plant coffee trees, spread fertilizer in banana plantations, weed or fertilize the cane fields, know that they are creating wealth. And wealth for whom?

For themselves!

On occasion we have spoken to coeds in the citrus groves and have asked them: "What are we going to do with so much citrus fruit?" and they say: "Export it. Give it to the people." For them, the answer to the question "What are we going to do with so much of this thing or that? is "Give it to the people." They have already begun to see that the people are the beneficiaries of all they do, of all they build, of all they create. They are beginning to see this clearly.

This new generation will develop, thinking in terms of creation rather than in terms of money, the intermediary between man and man's products. They will think money vile. They will think that for many reasons. Under capitalism, money was an obsession. Here, people call it "vile" money. The Americans call it "bill" money. They use the "b" and we use the "v" [*Laughter*]. Money is the intermediary between man and man's products.

That new generation grows in other ways too, and increased production facilitates all this. So, it is possible for a small farmer to be working in communism. Yes. If anyone was wondering "How will I live in a communist society?" the answer would be, the same as now, with one difference. He wouldn't have to pay for anything. And then you will say: "Well if they don't charge me, what am I going to do with these papers?" What would you logically do? Turn your things in and receive what you need. One goes to buy cigarettes, matches, clothes, shoes. Some will say: "Will there be enough clothes to meet everyone's needs? The needs, yes. Perhaps, also, the man or woman in charge of distribution will say, "Don't hand out any more clothes gratis, because all the women want 25 dresses each."

Then, organize a campaign, because this also has to do with the people's upbringing. But there will be clothes, even 25 dresses for each woman, if necessary.

Recently we saw several types of clothing of fine quality, made from synthetic fiber produced from cane bagasse. Could you have imagined this country's bagasse converted into fiber? Can you imagine the amount of clothes? You will say "Well, and if I want to wear a wool shirt and there are no sheep here?" My friend, we will send synthetic fibers where there is wool, so that they will send us wool, because they may have sheep, but they don't have bagasse. [*Applause*]

Notwithstanding the fact that we don't need much wool here because this is quite a warm country. But what I want to say is that the technical possibilities are incredible, incredible! And the possibilities for production are limitless. What does this require? That the people work, that everyone work.

Many women have gone to work with the opening of new nursery schools and the appearance of job fields suitable for them. In the future the entire population will work except the old people, of course, those who are ill and very young children. And with our entire population busily producing, aided by technology, we shall be capable of producing everything that man needs, and even much more than man needs.

For example, our country expects to build 100,000 homes a year beginning in 1970. [*Applause*]. One hundred thousand homes beginning in 1970! Will these be sold? No. Who will get them? Those who need them. Will they be charged a penny? No. Not even if they are worth 5,000 pesos? No. The residential areas built on State farms and near factories are already rent-free. Water—almost nobody pays for it, and I believe that the Hydraulics Institute has already decided to discontinue all charges for water everywhere. Electricity is another thing that one day the people will receive free.

One hundred thousand houses a year! Before, in order to have a house a man had to win the lottery. Candado soap raffled off a house every month and people had to win the Candado soap raffle in order to get a house. Everyone dreamed of having a house, a roof. "A roof, all I want is a roof to be happy, to have peace of mind!", etc., etc.

Every year from 1970 on, 100,000 families will receive houses

Gentlemen, the Revolution has been building homes,—not many, unfortunately. Some 7, 8, 10 thousand. That's not so many, but no one has won them in a raffle. Beginning in 1970, 100 thousand families will receive homes every year; in 10 years that will be one million roofs. What is a house? It is an object of value that man creates with his work, that is created by those who produce cement and steel rods, by those who transport the materials and those who construct the buildings, and this too, will also be mechanized more and more. You may ask, "and who is going to build 100 thousand houses? Our construction workers and machines, because prefabrication methods will be used in great part. Prefabricated houses, produced industrially. If we had to put brick on brick in each one of the these houses, nobody would get a house here—not even by buying Candado soap. [*Laughter*] I should say, not even in a raffle, because the solution to the problem lies precisely in the use of machines and technology.

And so our country will move ahead, solving all of these problems with the incorporation of all the people into our labor force and with the use of technology.

I have wished, comrade farmers, in this third Congress, to devote a part of the time to discussing these questions that I suppose will be of interest to all of you. [*Applause*] Nevertheless, today's problems remain. What do we expect of you, and what is being done to achieve this? The land you own must produce, produce to the maximum! [*Applause*]

What are we doing to assure that you, with the use of machinery—to the extent that this is possible—and with technology, obtain maximum productivity from your work and maximum productivity from your land?

There are some very illustrative examples: The coffee plantations in the mountains were producing some 50 to 60 hundredweights per **caballería**. A little more than a year ago, the plan for the introduction of technology in coffee production was begun, with massive application of fertilization, pruning and replanting of coffee bushes, the utilization of more productive plant varieties—in short, intensive work was undertaken in coffee raising. Today, 27,000 small coffee producers in the province of Oriente are applying technology, and the production of coffee is increasing rapidly—not as a result of new plantations, but with the old plantations, which include between 11 and 12 thousand **caballerías** of land. To produce 50 or 60 cwt. of coffee per **caballería** is shameful when it is perfectly feasible to produce 200 cwt. per **caballería**, which is only some 20 cwt. per **caró**—as the farmers call it.

Naturally, when they began to apply fertilizer, the farmers had a short phrase for it: "The plantations are like new." We have talked with hundreds of farmers, asking about the effect of fertilizers on coffee. They said: "Look, these coffee plantations were very old, 15 years old; they were worn out, and now they are like new."

That is, by spreading fertilizer, the farmers have seen their coffee groves take on new life. And on our last trip through Oriente Province, an up-to-date farmer who lives near San Lorenzo de Céspedes—way up there in the Sierra Maestra—told me, "Look, you said that we were going to reach two million cwt. in 1970" I said, "Yes." "But we're going to reach that in 1969, we're going to make it in 1969." "Is that so? Why?" "Look, I was producing 300 cwt. by 1969 I'm going to produce a lot more and by 1969 I'm going to reach 800 cwt." "Is that so?" "Yes."

He had seen the results of fertilizing the coffee groves, the increased production, how the coffee groves flourished, how they retained their flowers, their beans, everything. And that farmer stated the truth: The goal of two million cwt. by 1970 will be surpassed by a wide margin.

We'll reap two million cwt. in the old coffee groves alone. And some farmers have realized this. One thing is particularly interesting: Nearly 100 percent of the farmers are already spreading fertilizer. Every time one of these techniques is introduced, there are always some more advanced farmers and they begin to put it to the test; there are other farmers who are more cautious. They wait to see how it comes out. And of course, the effects have been so incredible that right now 100 percent of the farmers in the mountains are fertilizing the coffee.

And something along the same lines is occurring with tobacco.

Farmers should specialize in one crop

There are a series of principles that should be applied to small farming. First, it's necessary for farmers, everywhere, not to produce everything. Let us explain to you what we mean. One of the terrible things you see when you travel through the countryside is a lack of specialization by the farmers. And the farmers ought to specialize in one, two or three crops—but specially in a single crop.

What does this mean? Some farmers should be mainly tobacco producers. These are the farmers who for almost a lifetime have grown tobacco in the valleys of Pinar del Río, in Santa Clara, in different places. Other farmers ought to be mainly coffee producers. Others ought to be livestock raisers; others should grow root vegetables; others should be fruit producers. That is, an effort must be made so that the farmer doesn't produce a potpourri of crops—you, over there, you can produce potatoes—everyone should specialize.

Sometimes the farmer is not to blame for that lack of specialization that can be observed; we're partly to blame. It's true that there are special zones where coffee, for instance, has always been produced and will continue to be produced, but there was also another special zone, for example, in Oriente, in which the farmers used to produce beans—they were the farmers of Velasco, one of the zones with the best climatic conditions for bean production. Suddenly there was a shortage of papaya, and the price of papaya shot up. The collection center suddenly set such and such a price for papaya. The farmers in Velasco began to do some figuring: "Beans are a lot of work and bring me a lot of problems. If I plant papaya I'll make ten times more."

And the farmers in Velasco began to give up beans and start raising papaya. And what kind of deal was that for the country? We had to use our dollar reserves to import beans, because those farmers began to produce papaya instead.

Gentlemen, papaya is a very easy crop to produce. A little bit of fertilizer and irrigation and a good variety of papaya can be grown. On 100 **caballerías** (1,340 hectares) more papaya can be produced than the papaya produced by 10,000 farmers on small patches of papaya, because even to pick up the fruit along the way, trucks would need to make more stops than a local trolley: to pick up a hundredweight of papaya here, another there . . . and further along a man with a receipt, a contract, this, that and the other. [*Applause*]

We have suggested to a number of farm groups: "Produce papaya here." Because there are some crops that have been incorrectly assigned to the farmers, a number of crops that have been encouraged through astronomical prices. There is a need for cooking tomatoes, and up goes the price of cooking tomatoes. Well, there are some farmers who have always produced cooking tomatoes. Very well, those who have produced them have the right to continue producing them. But suddenly, some farmers who produced cane saw that their neighbors, with two or three cordeles, (1 **cordel** = 414.2 sq. meters of land), and with much less work, got much more money than they did with 60 cordeles of cane. Then, it didn't matter that a sugar mill was being enlarged, that the railway lines would be there, that everything would be there; suddenly the farmer said, "No, no, I'll dig up the cane that gives me such a headache and I'll plant cooking tomatoes".

Naturally, if everybody who wanted to, set about to plant cooking tomatoes, then there would be a surplus and they would not be worth a cent. Those are the methods of capitalism. How do they work under capitalism? "There is a surplus of this. I'm going to produce what is short." Everybody goes crazy trying to produce the scarce product. There is then a surplus of that product, and they abandon it, and set about producing something else. We cannot produce under the same conditions as capitalists.

We must say that we can't place the blame on the collection centers, but the price policy that they have been following is incorrect. [*Applause*] There is a shortage of carrots? They speak with two or three farm groups and these produce enough to make a mountain of carrots.

Don't encourage carrots here; there, cooking tomatoes; there, beets; there, papaya, gentlemen; because that creates an infinity of upsets, of problems, of headaches. The farmer who has traditionally been producing a crop, one to which he is accustomed, will, logically, drop what he is doing, because he considers that he is working much more and is earning much less than the others.

In short, the farmer who has traditionally planted tobacco should continue his cultivation of tobacco, improving his technique, raising his productivity; the one who has been cultivating coffee should continue cultivating coffee; he who has produced silage, who has cattle, should continue developing his herds; he who raises root vegetables should continue cultivating root vegetables; if rice, rice; potatoes, potatoes. The one who has produced cane has another problem. There is a problem with cane. For that reason I have not included it,—not that I have forgotten it, let me tell you. But I have mentioned these areas in which there are no problems.

The Velasco farmers have done very well. They were spoken to, they have been granted facilities, and they are all producing beans again. They are all excited about bean production and the struggle to raise the yield per land unit.

Naturally, it is not enough to establish a correct price policy, to give a guideline. It is not enough to say to the farmers of Velasco: "No, stop growing papaya, start producing beans again." No, you have to go there and speak with them, tell them: "What do you need, how many machines? Do you have fertilizer? We are going to apply fertilizer; irrigation can be applied so that you can produce during the dry season as well; we'll build a dam here, we'll look for irrigation equipment." You have to go and provide them with the resources.

And so we think that the tobacco growers should struggle to apply organic fertilizer to use irrigation. In tobacco there is a whole program of building small irrigation works to raise tobacco production considerably, doubling or tripling the yields through the use of organic and other, fertilizers and irrigation. There is a whole plan for tobacco, as there is for coffee. And this is what we have to do with all of the country's main crops.

Farmers in the mountains will be supplied with everything they need

Of course the farmer wants to produce something else for his own consumption. That's fine: If a farmer who grows coffee also wants to produce for his own consumption, that's fine. I assure you, judging by what I have seen in the countryside, that with a small amount of well-tended land, using fertilizers, any farmer can produce all he needs for his household, all that the family needs. That is, each farmer can have a main crop—coffee, tobacco, cattle, root vegetables—and at the same time devote part of his land to producing what he wants for his own consumption, if he so wishes.

And the commercial crops, one main crop, and then others for family consumption. Often farmers use too large an area for consumption crops because of their low yields; frequently four acres are used to provide for household needs, when the problem could be solved by applying fertilizer to one acre. Naturally, this must be planned in accordance, with the policy that is being followed by the Revolution of distributing fertilizers, of increasing the use of fertilizer for all crops. At times, in a plantain region, for instance, there are about 6,000 hectares of plantains, and then suddenly you come upon some brush. "What's that brush doing there?" "That belongs to a small farmer." Or you find a real mixture of crops: root vegetables here, maize there, plantains, a pig here, a cow there. Now, if the whole region is good for plantain growing, why doesn't that farmer grow plantains, and leave some land for crops for family consumption; let him sow root vegetables if he wants to, maize if he wants. But sow mainly plantains, so that when the airplanes come around to spray or to fertilize, they can spray and fertilize his plantains too. Sometimes I've travelled around with some **compañeros** of the Farm Group, and I've had to say to them, "Listen, this land is abandoned, raise plantains here. Call the farmer and tell him, "Listen, those plantains you see there are yours, take care of them, tend them.'" For at times you find abandoned parcels of land—because the owners don't have the necessary resources or because they make their living from other work or because of other such problems.

That is, we must try to get the farmers to specialize in certain crops. In some things there has been great progress, for instance with coffee. In the mountains, our policy will be different: We want to supply the farmers with everything they need through a roads system now under construction, so they will no longer have to raise plantains, maize, and a host of other things in the mountain zones.

What method of planting do they use in mountain farming. Without exception, on a slope they will plow up and down, never in contour. I have never seen a single exception, not one farmer in the mountains doing otherwise. They do not use contour plowing for plantain fields. It is always up and down. I have asked them, "Why do you plant this way?" Some have answered, "Well, it's easier for moving about." Others say, "If I don't plant this way, the rain will wash away the seeds" or "It's very hard to hoe the field any other way" or they give this, that or the other explanation. No matter what the reason may be, every single mountain farmer plants by following the slant of the slope.

The State Department of collection centers is busy there in some of the mountain areas. When we came down from a region called Punta de Lanza, we saw a few plantain fields here and there, and we asked, "What are they doing here? Is this a commercial crop?" They answer was, "Yes." "To whom do you sell?" we asked, and the reply was, "To the collection centers."

Is it correct to grow plantains in these mountains—where scarcely any topsoil was left following the onslaught of Hurricane Flora—when thousands upon thousands of hectares of land are being planted to plantains in the flat lands, with adequate irrigation? In the Cauto Valley alone there are 6,700 hectares of plantain fields, and by next year there will be 26,800 hectares. Yes, 26,800 hectares of plantain fields in the Cauto Valley! [*Applause*] It will be very easy to supply the farmers in the mountains with all the plantains they want and ask them to grow coffee instead of plantains.

This means that a correct policy has not been followed by fostering plantain growing in mountain zones. We have spoken to farm groups and have asked them to set apart land for growing cassava. There's a belief that cassava can only grow on mountain lands, but it can grow perfectly well on flat land. We are looking for such lands so we don't have to ask farmers to plant cassava in the mountains.

So in the mountain zones we are going to follow a policy of supplying the farmer with everything he needs, including root vegetables, so that he may concentrate on coffee growing. Now that the coffee is being planted in rows we advise that gandul beans be planted between the rows. These beans should be raised not as a commercial crop but rather as a plant to protect the soil from erosion. Besides, this crop is suitable for human consumption or for use as chicken or hog feed.

Maize is not the only thing that's good for chicken feed. If the contour method is used on the new coffee plantations and the rows are three meters apart, the gandul bean can be planted in between. It can be fertilizer and will serve for hog and chicken feed. Moreover, it is good to eat, especially when fresh, and many farmers cook it with rice. That is a correct solution: not to grow maize in the mountains but rather grow some other crop that will solve a problem, be easier to cultivate and not harm the land.

As I told you at the beginning, very little was known of agricultural methods.

I'll tell you something else; with a few exceptions . . . there are farmers who are quite handy . . . On the border of Camagüey and Las Villas, on the banks of the Jatibonico River, we met a farmer who impressed us. He had a small citrus grove; he had planted the trees himself; pruned them; and was getting a magnificent yield. Is he around here some place? [*Answers of yes*] Well, I think his name was Mesa. That farmer had planted cedars, citrus groves and pasturage there; he was making a home-built turbine there, going through all sorts of tribulations looking for a motor. He searched for fertilizers and there were none appropriate for citrus trees, but he got another kind and was adapting it with real skill and technique. That farmer was a real agricultural expert.

At the entrance to La Montería, in the Sierra Maestra, we came across the unusual case of a farmer who had planted a mahogany tree, a baría tree, a cedar tree, a whole series of hardwood trees . . . Is he around here too? [*Replies of "Yes"*] Well, I'm delighted! These men deserve to be here because they are truly vanguard farmers. That farmer, Verdecía, had planted dozens of trees. He had done something to make up for the sorry sight of so many forests destroyed with never a new tree planted.

I asked him, "Do you think the other farmers will plant trees too if they are supplied with the seedlings?" He answered, "Yes, of course they will." This man has been elected "model farmer" of that region. He has planted hardwood trees everywhere he doesn't have coffee growing. By now there's a veritable fad of hardwood planting. The comrades of the Manzanillo zone wanted to expand the plan but I said, "No, hold it!" Every once in a while you have to say "hold it"—there is no lack of enthusiasm here, there's plenty and some to spare. What's missing is correct orientation [*Applause*]. And the lack of orientation is not lack of good faith. It's the result of all the ignorance that has accumulated in this country.

We plan to set up nurseries throughout the Sierra that will supply 200 to 300 million trees for timber

I'm going to give you an example. Today, when we were talking with a group of comrades who are developing the cultivation of pineapples, we asked them about the seed, because the increase in pineapple production is limited by the amount of seeds, and they told us that a new technique for getting seed has appeared. I asked: "What's that?" They said: "The stalk of the pineapple is cut crosswise, like this, and each stalk produces at least 12 shoots." I said: "How's that?" And they even showed me in one place a plot where they had the shoots that had been obtained from the pineapple stalks, shoots that had some leaf-buds already growing". "And no one here knew that?" "No." "And where did this appear?" "In a mechanics magazine that came from I don't know what country." I said: "Can this be possible! In this country where there were at least a few people that were said to be agricultural technicians, agronomists, and no one knew that?" Well, that's a fact. No one knew it. Such a simple thing, something that was ten times simpler than the famous "Egg of Columbus." The pineapple stem could reproduce pineapples, and nobody knew it.

It's incredible the number of things that we didn't know,—that we still don't know, for that matter. That's the reason that orientation doesn't exist. Correct orientation is seldom provided through ignorance.

But in the zone of Jíbaro and Montería we had to tell the comrades: Wait! Why did we tell them to wait? Simply because we expect to carry out a forestation plan in the Sierra Maestra in 1969, a vast plan: We plan to set up nurseries throughout the Sierra of 200 to 300 million hardwood trees and provide them for the farmers who want to plant them; but now, in 1967 and 1968, we are working with coffee; we're introducing technology in the production of coffee, re-seeding the coffee groves, developing the coffee groves. If in the midst of this coffee plan that we're carrying out in 1967 and 1968 we try to introduce a forestation plan, we're going to tangle one thing up with the other. And I suggested to them that they continue in that region as a pilot plan to see what the farmers' attitude is, how they accept it, how enthusiastic they become. Our aim is to give the farmers free seedlings, and also the fertilizers for the forest plantings, free.

With trees you have to wait 10 or 12 years for them to begin to produce, but no more than that. We believe that a mahogany, a cedar, any of those trees, can be cut in 12 years if they are fertilized. We are also going to use fertilizer on the forests.

Also, we are going to say to the farmers, "There you have a bare piece of land that isn't producing a thing, where there is no coffee, here you have the seedlings in their little sack and all. Plant it, here is some fertilizer." Every year. If, later on, he wants to exploit that wood, he sells it; if someday he wants to sell his piece of land, it will be worth much more, although I think that the argument of improved value is not the one that is going to convince people very much here, because, as I said before, as the years go by, money will be worth less, for the reasons I explained to you previously.

But of course, whether an individual is interested in money or something else, what interests the country is that there be not any fallow plots of land, the country is interested in planting trees. On State lands there is no problem: in the mountains we are going to go to all the pastures and all over leaving certain areas fertilized with a view to milk production, and we are going to sow the rest with hardwood trees, or pines, according to the characteristics of the soil and terrain. Soon we are going to carry out a plan in the Sierra Maestra—starting in 1969. We are also going to carry it to the Escambray and other mountainous regions. We cannot do it before that because now we will be working on coffee.

That is, there are farmers who are interested in technique, who are concerned with improving the land, who are concerned with planting trees. There are farmers like that. And it is really very encouraging every time we meet this kind of farmer.

Outside of that, I'm going to tell you something: the level of agricultural know-how among the small farmers is more or less the same as when Diego Velázquez began the colonization of this country; our agriculture, our small-scale agriculture, with some exceptions, has the same technical level as four centuries ago, four centuries ago! Of course, here, too, there were always some crops that used to be irrigated and fertilized: potatoes, some tobacco, a little rice, but, outside of that, cattle, most of our cane, root vegetables, all the rest of our agriculture was anachronistic, anachronistic!

It never occurred to anyone to fertilize bananas. In Oriente there are banana plantations that are 20 or 30 years old that have never received a pound of fertilizer.

When I tell you this, I am not blaming you for it. I am simply referring to a reality, and it is the reality of our conditions in the past. There was no technical knowledge, there was no fertilizer, there were no markets or credits, there wasn't anything. I mean that our agriculture is very backward, our small-scale farming is very backward.

Tell me something, some banana-growing-farmers—there must be several here. Is there a banana planter of Holguín, of Banes, of. . . ?

[*Someone in the crowd answers affirmatively*]

I would like to ask a question: How much land do you have planted to bananas?

"16 strips."

And what is the greatest number of cwt. you have produced?

"1,700 cwt."

That is plantain. Of the fruit variety, how much do you think can be obtained from one **caballería** of land? Are there any fruit banana growers around?

—"3,000 plants."

How many bunches? How many cwt. would it come to? Can 2,000 cwt. of fruit banana be obtained from one **caballería**?

—"Two thousand cwt., more or less, can be produced on two strips of land."

Do you know that, for example—and according to our estimates of the results of some small experimental plantations—that as many as 20,000 cwt. of fruit banana can be produced on one **caballería**?

We mean to make maximum resources available to farmers to help them raise productivity

If we had not been very busy recently, we would have been able to suggest that the Congress delegates visit some of the experiments being carried out. [*Applause*] Banana plants that start to bear fruit at five months and before the year is out are producing bunches of 70, up to 85 lbs. Up to 40,000 bunches can be produced on one **caballería** of land, and the plants are still an adequate distance apart.

Now I'll give you the example of another crop: coffee. People are picking around 50 or 60 cwt. per **caballería**. On some of the plantations we are setting up we plan to harvest up to 1,000 cwt. per **caballería**.

That is, there are some plantations in Cuba that require certain technique, and when those techniques exist it is not too easy to duplicate them, but with most Cuban crops, we can triple, quadruple and quintuple production—even increase it tenfold.

We propose to provide—and in the Revolution there is now possibility of doing this—the maximum resources for the farmers so that they may increase production—double, triple, quadruple it. [*Applause*] For example, right now, banana fertilizer is being delivered to all the small farmers [*Applause*]; coffee fertilizer is being delivered to all the small farmers.

Four tons of fertilizer per **caballería** have been distributed among all the cane-growers, and, in addition, they will be given one and a half to two tons of ammonium nitrate per **caballería**. [*Applause*]

All the tobacco plantations will be fertilized also, and will be irrigated as thoroughly as possible by building small dams.

Fertilizers have also been distributed among all the small farmers who grow root vegetables, beans, fruits, and citrus fruits. That is, we are already on the way to applying technology.

Artificial insemination is being introduced on small cattle farms

Artificial insemination is beginning to be introduced for the cattle of small farmers. Cattle-raising is one of the sectors of agriculture of the ANAP where some of the best work must be done. As I was saying, artificial insemination is already beginning to be introduced.

You tour this country and you see lean cows all over the place. Poor little lean cow, and they are milking it . . . I don't know what they're extracting from it, maybe its very life. All over the place you see cows tied up, cows in a pasture without feed. And the truth is, when you have a bit of awareness of the care that must be given to animals, you suffer when you see that.

There are very few farmers that have cultivated pastures. Almost all of them have natural pastures. In general, not one pasture belonging to small farmers has ever been fertilized. And there are many hundreds of thousands of hectares of land devoted to cattle. We aim to promote cattle raising just as we are promoting coffee, tobacco, root vegetables, all the other crops.

What policy do we plan to follow with cattle? Naturally, in the zones near Havana and other large cities, we are going to develop dairy cattle, but in general in the provinces of Las Villas, Camagüey, and Oriente, we are going to propose three things to the farmers: first, sowing artificial pastures, including leguminous plants with the ordinary grass; second, fertilizing of the pastures; third, guide the small farmer to raise beef cattle rather than dairy cattle.

Why? Except in the western regions—where the farmers have a dairy tradition, where there is a large population concentration—in the rest of the island, it is much easier for a farmer to have a herd of beef cattle than a herd of dairy cattle. Dairy cattle are a lot of work, a lot of headaches, the daily milking for example; it is much easier for a farmer who has 15 hectares of pasture, of land for cattle, to sow it with artificial pasture, to fertilize it, to keep 30 or 40 cows there and raise 25 to 30 calves a year. Just by giving them a minimum of care, by moving them from one part of the pasture to another, practically just looking at them—just looking at them—a farmer can tend a herd of 30 to 40 cows; that would mean a good income for the farmer, and an income that would not give him many problems. Milk is more work, and can be produced on the State farms.

Take, for example, the Bayamo region. When the whole plan for the Bayamo region is developed, the State farms there will produce a million and a half liters of milk every day. Collecting that milk will be easy, because it will be from a few hundred dairy farms. If the same amount of milk were produced by small farmers, you would have to go to 10,000 different places every day to collect it, pick up 10 liters here, 15 over there, 20 elsewhere. That milk would be produced under 10,000 different conditions of hygiene. It can perfectly well be produced on the State farms.

That doesn't mean that if a Bayamo farmer is a dairy farmer and wants to continue, that we won't help him. We do help him, we'll supply his cows with artificial insemination, provide resources, everything. We just . . .

[*Someone in the audience says something to Fidel*]
Where are you from?
—"From Las Villas."
You're from Las Villas. What municipality?
—"From the Taguasco sugar mill."
Next to the town?
—"Yes."
How much pasture land do you have?
—. . .
How many cows?
—. . .

How many do you think you could have there?
—. . .
On 20 hectares?
—. . .
When you have artificial pasture . . . you already have 13 hectares of pangola grass. Do you have it divided up into sections?
—"Yes."
How many?
—
Have you ever used fertilizer?
—"Yes."
When?
—. . .
Ah the thing is, you're a model farmer. I'm not talking about cases like yours.
[*Laughter and applause*]
That means, you're going to have calves of Brown Swiss or Holstein.
—"Holstein."
Do you want to be a dairy farmer, or what?
—"I like milk production."
How many cows can you milk every day?
—"25."
You have to milk 25 cows every day, that's hard work. That is, if you had, instead of dairy cattle . . . I'm not saying that you should give up dairy cattle. I don't know the concrete situation of the town of Taguasco, but what I mean is, that on your hectares, you can have from 40 to 50 head. If instead of inseminating these cattle with Holstein, you do it with Charollais or Saint Gertrudes, by the second or third generation, you already have wonderful beef cattle. On about six hectares you have ten sections, fertilizer, and the only thing you have to do is move the cattle from one section to another every four or five days. And do you want me to tell you how much you would produce? You can produce seven or eight thousand pesos in foreign exchange every year on your 20 hectares very easily.

We will suggest that state farms produce milk and small farmers raise cattle

Look, what I'm proposing here we still cannot begin to do. Why not? Because the great milk increases as a result of present efforts will still not come about until at least 1970. In 1967, 1968 and 1969, we will still need the milk production of private farming in Las Villas, Camagüey and Oriente; in Matanzas and Havana, we will always need it because here a good part of the pasture land belongs to small farmers and there is a very large concentration of population. What we are going to suggest is that the State farms produce milk and the small farmers breed cattle. Because, for example, in a place like Camagüey, where there will be from two and a half to three million cows in the future, one million will be devoted to milk production, another million to breeding and the rest of the cattle of the private sector can be devoted to beef production. We suggest that the State farms produce the milk Why? Because labor supplies are required, refrigeration is required, equipment is required to preserve milk, and the collection of milk must be well organized. It is easier to pick up the milk if there are 100 dairy farms producing 1,000 liters each. The trucks make the rounds of those 100 farms, the milk is kept cold in those 100 places. It's a far more difficult problem to preserve and collect milk in places where production is 50, 70 or 100 liters.

How many liters do you produce a day, for example?

—"Seventeen."

As a good farmer, imagine the problem of picking up 100,000 liters of milk. Some 5,000 small farmers like you would have to be visited daily. To keep this milk clean and refrigerated during the collection, 20,000 refrigeration units would be required. Whereas 100 places could do just as well.

However, if you were raising beef cattle, you would only need to turn in your product once a year, when the dry season began, or at any time; only one trip would be required to pick up the product.

And since it is much easier to produce meat than milk, in the future we plan to aid all farmers in planting their meadows and brushland with cultivated grasses, **pangola** with **kudzú**, which is what your particular farm needs. On your seven hectares, you should alternate rows of **kudzú** and of **pangola**, so that you do not need to apply nitrogen; this will improve your cattle feed greatly.

Is Evidio around? That's a shame, because Evidio is producing 110 liters of milk with eight cows—with eight heifers which are giving an average of 13 liters of milk a day using guinea grass with **kudzú** as feed in this case. Evidio is a farmer from San Andrés.

Certainly if a farmer wants to raise dairy cattle he should go ahead and do so. But we suggest that, in general, farmers raise beef cattle instead.

And next year, don't worry, you can have your cow inseminated from a Charollais or St. Gertrude, stud bull, whichever of the two beef cattle you prefer. In the province of Las Villas we'll probably use either. St. Gertrude or Charollais for insemination. Those are full beef cattle; the second generation of offspring (F-2) are already three-fourths beef cattle.

Your cows are zebu. In four years the nature of our cattle will be transformed; it is likely that we will use the Charollais as one of our principal beef cattle, since it is stock of such high quality and value. Those of our farmers, from Las Villas to Oriente, who raise cattle will have real beef stock.

[*Something is said from the audience*]

Insemination of dairy cows is a correct procedure. We should not alter this now. This year we are going to inseminate with dairy studs, because that work was begun, because the idea is relatively recent. That is why the artificial inseminators are going to keep on working in Sancti Spiritus and in Bayamo. [*Applause*] But that doesn't matter.

Because, for example, if next year your region has the Holstein F-l with zebu, that Holstein F-l with zebu can be inseminated with a beef steer and it will be easier to raise than the zebu. It changes, because a milk cow, if it has good pasturage, feeds its calf much better than the zebu. Without a doubt, that F-l can feed its calf and reach the age of one year weighing 700 or 800 pounds easily, as long as it has grass.

What I'm saying now is a general guideline for the future; that is, we must ask this of the cattle raisers, because this is precisely what they like . . . Cattlemen like cattle raising more than milk production, and we have concluded that, if they like to produce meat, if it's less work for them, if, moreover, collecting the milk is much more difficult than collecting meat, we should suggest that they dedicate themselves to raising cattle, leaving milk production to the State farms.

The policy that is to be followed is that of specialization of production

By 1971 we shall be able to say to any farmer from the limits of Matanzas and Las Villas to Baracoa:

"If you like, concentrate one hundred per cent on meat production." Of course, around Santiago de Cuba, Cienfuegos, the city of Camagüey, the large cities, it is logical that milk production be continued, but even in places such as Bayamo an important milk area—State production will be about one and a half million liters—that is, the farmers will have the choice of being cattle-breeders or milk producers, according to their preference. This is how we're going to proceed.

Next year, in 1968, we are going to start stimulating the cultivation of pasturage in Bayamo in the private sector; we plan to plant 1,000 **caballerías** of pasturage for the Bayamo ranchers. But we are not yet going to establish this plan throughout the country. In zones of Havana Province, yes; in this zone. By 1969, we plan to give a tremendous boost to the planting of pasturage in the private sector. And, of course, we can begin insemination with beef studs next year, since the offspring won't appear until 1969 and will not be in production until 1971 or 1972. And by that time the production of milk is going to be enormous, so that it will be more important for farmers to dedicate their pasturage to the production of meat.

This, then, is the general guideline: some farmers will be producing mainly tobacco, others root vegetables, others beef cattle, others beans, others rice. We are also going to give a great boost to rice production in the regions of Manzanillo and Bayamo; a plan will be made with the small farmers for irrigation, fertilization, everything in that region. The policy we are going to follow is that of specialization in production.

This is not new. This specialization of production is nothing new; it has been the traditional development; what I want to say is that we shouldn't be changing from one product to another, from one thing to the other. If a product does not offer satisfactory results to a farmer, there can always be a solution in regard to that product.

For example, there is the problem with the cane that I haven't mentioned yet. This problem is more complex. There are provinces, such as Matanzas, Las Villas, Havana, where there are a great many small farmers growing cane. We can't tell them: "Well, if cane is a big nuisance, stop producing cane." We can't tell them to stop producing cane, because the sugar mills are there, the installations, everything.

Why? Because they own a great part of the cane lands. We even have to struggle, to insist, because there has been a policy that goes against the national interests. What is that policy? Ripping up the cane fields. Ripping up the canefields harms the economy; it is preferable to look for another solution—anything but ripping up the canefields. We have no other alternative in those provinces where the small farmers have a lot of cane than to ask them to keep on growing cane, and try to find a way to compensate them so that they may feel satisfied with what they are doing.

In some areas, such as Camagüey, where the small farmers have very little cane, they showed me a project for planting the cane, except in the case of lands very close to the sugar mills, on State lands, so that the farmers could concentrate on raising cattle. There is a cattle-breeding tradition in Camagüey and the small farmers have very little land devoted to cane; there is enough State land to permit the transfer of these cane plantations to State land—and thus a solution to the cane problem in Camagüey can be sought.

But, nevertheless, in Las Villas, Havana, Pinar del Río and Matanzas, the same thing cannot be done because there is not enough land. The sugar mills are there, and it is impossible to do without cane-growing. What has to be done is to mechanize, send in cane-loading machines, achieve maximum productivity.

From force of habit, an incredible thing has been happening here. According to the comrades of the ANAP, the only farmers who pay taxes in this country are the cane-growers, because even the coffe-growers, who paid some taxes, were exempted.

Therefore, the Government has drafted a law whereby the cane-growers will be exempted from taxes. The fact is that cane is extremely important for the economy: it produces sugar, molasses, bagasse, cogollor (which is fodder for cattle); cane is extremely important, and it is also difficult to cultivate, it is not as easy as other kinds of cultivation, and it turns out that the only tax-paying farmers here are the cane-growers! We are going to lay the blame on the ANAP! We have to blame somebody, so we're going to blame the ANAP.

Pepe (Ed. note: ANAP President) is the one who has been insisting on this problem.

As of this year small cane growers will not be paying any revenue tax on their cane

Gentlemen, the culprits are the cane-growers, the cane-growers themselves. Because they continued to pay I don't know what kind of tax and practically all taxes have been eliminated; they kept on paying, because it was a discount arranged through the sugar mill, according to what they have been explaining to me . . .

What happened? What happened to the small cane-growers?

[*A delegate speaks from the floor*]

This year? This year, then . . . [*Someone says: "Last year"*] No, no, this year no tax will be charged. [*Applause*]

Of course, these taxes remained because they were upheld by law. If this problem had been raised before, it would have been solved. Now, as I was saying, from this year on, that tax is suppressed, that is, you will not have to pay any tax on the cane you have cut this year. [*Applause*].

This has to do directly with what I was telling you about incorrect directives; this was one of those incorrect things that still persist. Here we are working to raise our cane production, in order to reach a harvest of 10 million tons, and we find that the small cane-growers, responsible for almost 25 per cent of our cane production, were paying taxes. Incredible, isn't it? Naturally, they were taxes that were already on the books. And what used to be paid? Formerly, rent was paid in addition to the tax. Rent payment was abolished, but the tax stayed on . . .

What do you pay, comrades? Tell me, how much are your taxes?

[*Reply from the floor*]

Social security is included in that 11 per cent?

[*A farmer answers: "Fidel, we don't have the rights to social security."*]

You don't have rights to social security. . . ?

PEPE RAMIREZ: The law establishes the option . . .

FIDEL CASTRO: What option? What is the option? Come up here, Pepe, and explain this.

PEPE RAMIREZ: [*At the microphone*] The problem is that, according to the Sugar Coordination Law, they had to register for it—the so-called option. Those who hadn't done it before didn't have their rights established by the Law. Nevertheless, after the Revolution, they have been charged for it; it is discounted, but they are not beneficiaries.

That is what I was telling you, **Compañero** Fidel; that we had been talking with Risquet, Minister of Labor, about studying this and finding a solution, because there are some who pay it, who made the option, and others who pay it without having made the option.

FIDEL: I still don't understand. Pepe seems to be hoarse. Let us see about this. The banana-growers, for example—do they pay this 11 per cent discount? [*Voices shout "No"*] Is there any small farmer who pays this discount? [*Shouts of "Quite a few"*] Who are they? Because listen, there is a part that corresponds to the workers . . . [*Shouts of "This is the one that corresponds to the workers."*]

That is the workers' withholding tax.

Look, gentlemen, I am going to give you my opinion on this matter. When I began my speech here tonight I explained the problem of money, all that sort of thing. I explained how money was gradually going to become less and less important. I also explained the problems of production. Do you want me to tell you what those taxes mean, what the 11 and so on mean? Do you think it will be of any use? Do you know what really counts? To produce 20,000 cwt. per **caballería** instead of 10,000 per **caballería**. That's what counts. [*Applause*]

What do we gain by having you pay 100 pesos while there is a production loss of 1,000 pesos? What can the 100 pesos be traded for? What the country is interested in is the production of an additional 1,500 cwt. of sugar per **caballería** to be exported in order to import goods equivalent to the value of the additional 1,500 cwt. per **caballería**. What we are interested in is a production increase since, in our country, everything that is produced is distributed. Have you ever seen anybody eat a peso bill for breakfast? Had anybody ever breakfasted on fried eggs and a 10-peso bill? Not too long ago, we met a farmer in Gran Tierra who has the habit of eating glass. Well, he was so sick we had to take him to the hospital. He was a regular "glass eater". But we have yet to see anybody who eats "peso" bills. I haven't seen such a case anywhere.

That 11 per cent is not edible. It is possible to eat

the sugar produced by that **caballería** of land and it is possible to distribute the sugar. The whole thing is an anachronism. What is our opinion on this? We believe that all anachronisms must be eradicated, that they are not important. What we're interested in is cane. What we are interested in is to have each **caballería** produce twice as much as it ever produced before, not in collecting 11 per cent or anything like that.

All workers without exception should have the right to a pension

Therefore, in any sector of small farming . . . If you were a merchant I would say, "No, no, the best thing to do is to raise the tax to 25 per cent", but we are not interested in doing that to any farmer-producer who works his land. This involves the problem of social security, but it all depends on the concept of social security. If we now collect 100 or 200 pesos every year from a farmer in order to give him a pension when he retires, the fact is that within 20 years that farmer will not consume what he is producing now but rather what he will be producing 20 years from now.

We believe that all workers, without exception, should have the right to retire [*Applause*] without having to resort to a capitalist method. What is important is that the small farmers produce, and, in order to produce, they need modern agricultural techniques, machines, fertilizers, resources and orientation.

Our national economy is not at all interested in these taxes. Therefore, we will study the problem of retirement as a right of every farmer. Of course, there will always be a "but"—possibly the farmer who sells turkeys out on the highway at fifty pesos a head. What shall we do with him? [*Shouts*]

Gentlemen, we have said that, in the future, money will be meaningless. Even retirement pay will have no meaning. Will anybody be in need of a pension when everything he needs will be available to him gratis? [*Shouts of "No"*] Very well, but still there are some farmers who need a pension and we still haven't reached the time when things can be had gratis.

We believe that the solution is to give all the farmers who have been farmers, who have worked, who can show that they are honest, that they have not taken part in "deals" speculation, and things of that sort, reason for rejoicing. [*Shouts and applause*]

And those taxes, do you want to know what caused them. Bureaucracy! Red tape! Then the harvesting of pesos instead of sugar cane is the harvesting of papers, is the creating of papers, of red tape, and this is a drain on the economy because probably there will have to be 500 people looking after those papers, people who would be more useful on a coffee plantation filling sacks with coffee. [*Applause*] Then it is evident that the still existing anachronism is going to be overcome by all those discounts. Would the sugar cane farmers be satisfied if all those discounts were removed? [*Shouts of "Yes"*] Would the sugar cane farmers be willing to make a supreme effort to fertilize their crops and apply technology in production? [*Shouts of "Yes!"*] Resolving all that, can we count on your 100 per cent cooperation in the plan for 10 million tons of sugar in 1970? [*Applause and shouts of "Yes!"*]

[*They tell him: "Listen, Fidel, in spite of all the taxes, we, the workers and farmers, were ready to surpass the goal of 10 million tons." [Applause]*]

Good! I wanted to tell you, as we have touched on different types of farming here tonight, that perhaps there is a farmer who has always raised tomatoes or beets. We are not referring to specific cases of such farmers, nor do we want to say that we want farmers to stop planting beets; we don't want to affect anyone. What we do say is that we are not going to stimulate crops such as beets, carrots, papaya, etc., that were sold at very high prices and created real confusion in agriculture. Everyone should produce the crop which he is accustomed to producing, the one customarily produced in the region where he is located: tobacco, sugar cane, pasturage, coffee, vegetables, whatever it is; everyone should be satisfied with what he is doing; and everyone should receive as much aid as he needs in raising the productivity of his land and of his work. That is what we want. And to have a policy of prices that do not spiral . . . It is a bad system, one that is against our ideas, that goes against socialism when, because of the shortage of any product, its price is raised so that it will be produced—especially when it is a product which can be grown on a few **caballerías** of a farm.

That is what we would like to say: our hope is to have farmers specialize, making the crop which they like their principal crop, and to have them try to obtain the greatest yield possible from their land and from their work.

That is, we need to introduce a much better orientation in private farming. And, above all, we have to introduce technology—we have to introduce technology in coffee, in tobacco. Some are more developed than others, but in cattle raising, especially, we have to introduce technology. The latifundia employed very backward techniques in cattle raising; some of them had a few very pretty little animals which they used to take to the fairs; almost all of them used natural instead of artificial pasturage. Practically no one employed legumes; no one fertilized his land; no one used artificial insemination; no one tended his herds, that is, there was cattle raising for the fair, with some pretty little bulls. And we have to treat all the cattle raising of this country as if this country were all one big cattle fair from one end of the island to the other. [*Applause*]

Society grants loans to farmers to make it possible for them to work and produce

And the problem of scrawny cows and other hungry animals has to be solved in this country. The time must come when the sight of a thin cow will be as affecting as that of a thin child; because surely those are the cows that are producing the milk so that those children will not be thin. And it should hurt us to see animals suffering from hunger, because animals are not our enemies; they are our friends, and they help us by providing us with food. It is criminal to have a cow that is hungry; it is criminal. It is not correct, and it is not revolutionary. If the cows could protest and were organized I am sure that they would have made a great protest in this country. [*Laughter*] But I, at least, am appointing myself their lawyer. [*Laughter*]

[*Someone says something to Fidel*]

With tobacco? Do you pay a tax? Good grief! [*Shouts*] The interest on the credit? and how much interest do you pay on the credit? [*Replies shouted from the audience*]

Well. I see you want me to do away with everything. Shall we do away with private property once and for all? [*Shouts and laughter*]

I say that because . . . Look. Many products bring a better price. We have tried to help these small farmers to a maximum. Now then. I don't know about those discounts. But look. They tell me that the tobacco growers don't pay taxes. Then what other discounts are there? Interest on credits? I'm going to tell you something.

That's a lot of rubbish in the double sense of the word: in amount and ideologically.

We should not collect any interest, because that is capitalistic. I'm going to be frank with you. It's bureaucratic, and doesn't even amount to enough to pay the expenses that the paper work involves.

[*Someone says something to Fidel*]

You're right. We're going to study the problem of that tax that is being paid, that was 10, 8 or 7 per cent. We reduced it to practically nothing, but there was still interest, and that is a capitalist concept. Really. It is money earning money. What is important to society when it gives a farmer money is that he give himself to working and producing and not to complaining that he produces to pay a tax. That doesn't solve any economic problem. That is an anacronism and a relic of the past. [*Applause*]

That farmer is correct.

What is it?

[*Someone says something to Fidel*]

The bookkeeping? They tell me that that is the business of the Credits Coop, and that the problem of bookkeeping charges will be taken up with the Bank (National Bank) and nothing will be charged for it.

Look, gentlemen. What you say is true . . . But in these days we have been very busy, and have not been able to have more contact with the Congress. All these questions will be solved in light of the new ideas of the Revolution. The Revolution has grown, evolved, developed. The ideas of the Revolution have developed. And in the light of these ideas, a low interest tax that spawns bureaucracy and does not solve anything . . . in the light of these ideas, discounts like the 11 per cent one, and all such things in a sector like cane that is of such importance to the economy . . . in the light of these ideas that are increasingly revolutionary, of a Revolution that aspires one day even to replace money by offering to the people all that they need as today they are offered education, hospitals and all the other things—all those ideas seem like anacronisms; they seem really out of date.

What interests us in the light of our own ideas

is that a farmer, instead of producing 7,500 cwt. of cane per *caballería*, produce instead 15,000 or 20,000 cwt! What is of interest in the light of these new, really revolutionary ideas, is that a farmer produce 600 cwt. instead of 300, whenever possible; that instead of 2,000 cwt. of bananas he produce on the same land 4,000, 6,000 or even 10,000; that instead of 10 lean cows per *caballería*, he has 35 to 40 fat cows per *caballería*; that an orange tree produce 1,500 instead of 500 oranges, because this is what people eat, this is what is distributed, to grab off a peso and moreover fail to raise production, is deception. The farmers should obtain larger incomes by producing more, because by producing more the people will have more of all those things that the farmers can produce. I think anyone can understand that. [*Applause*]

We have to introduce new ideas into technology and discard the old ideas in our scheme of things

Because if the lean cows, the 10 lean cows produce six lean calves that spend three years in the pasture to get to weigh 1,000 pounds, if that farmer has 40 cows and produces 30 or 35 animals every year—or even up to 40—let's say from 30 to 35 animals that weigh at one year or at 18 months, 1,000 pounds, that farmer is producing six times as much meat or exporting it and the people are consuming six times as much to import other things that they need. I think that is very easy to understand. The thing is, we still have a lot of old ideas that influence people's thinking, and we have to introduce new ideas in techniques and eliminate old ideas from our way of looking at things.

I believe that you understand perfectly well what I mean. It is clear enough.

[*Affirmative exclamations*]

Who wanted to say something? Who was it? Just a moment, there's a disagreement, two people want to speak at the same time. Which is the younger of the two?

—I am, I am!

FIDEL: Well, then you're the last to speak . . .

[*Laughter*]

We're going to give the floor to the one farther away . . . No, that one farther back, he'll have to shout more . . . no, you don't have the floor, the other man over there has it.

[*They say something to Fidel*]

Well, gentlemen, are we going to hold another congress here in the small hours of the morning? No. You already had your congress, and all those problems have been discussed, the problem of the cane-loading machines, everything has been discussed in detail. Here, what we are interested in are the general guidelines.

But we're going to give the floor to the last one, to two more who may want to say something, so they don't feel . . .

[*They say something to Fidel*]

They say that the Agronomy Department does not have to approve the credits any more, they say that the local farmers' association works out the credits with the Bank. That's good, the farmers' organizations should play a larger part in handling light machinery, in the question of credits, all those things. We are going to be greatly helped to solve all those problems as there is greater participation by local farmers' organizations.

Tell me, **compañero** . . .

[*Members of the audience say something to Fidel*]

Dry or irrigated rice? And how much damage do the wild ducks do? Have you heard that there's a group of hunters killing wild ducks in Sancti Spiritus?

"I bagged nine . . ."

FIDEL: So you got nine?

How many wild ducks have they killed in the past few days? About 1,200. Wild ducks are, at present, the number one enemy of rice. But the problem will be solved this year.

Where are you people from?

. . .

FIDEL: Ah, I was around there the other day. If you tell me exactly, how to get there I'll pay you a visit to talk about rice. What's the name of the place.

"Agricultural Society."

FIDEL: Agricultural Society.

Fine. Speaking of technique, we also recommended to some Las Villas tobacco growers that instead of doing odd jobs all years, after they picked their crop, they sow black-eyed peas in spring so as not to leave the land bare; we suggestd that they try this. Those beans are good for both human and animal food, and they protect the soil and give it organic matter.

But precisely because of what I was saying to you before, we are going to begin with a small farmer, a few small farmers, so they see how this works out, so they see the results, so they see that far from being harmful to the soil, the beans help it and will better protect it in the spring, because uncovered lands are easily eroded. But I think that the farmers are used to planting and doing odd jobs all year round until the crop comes up again. And we're going to do that.

You produce 1,000 cwt. of rice; that's a small amount. We have to produce at least 1,500 there. Is that clear? Fine. In Las Delicias. I'm going to visit you soon at that Agricultural Society. Fine. No, no, no more turns to speak. It's one thirty in the morning. Well, are we agreed that this *compañero* is the last to have the floor. Your word of honor. All right, let's see.

[*Members of the audience say something to Fidel*]

No, no, no then you don't keep your word, because this is already becoming . . . You've given me your word of honor that no one else would ask for the floor.

[*Members of the audience say something to Fidel*]

Oh, wait a minute, the one we agreed with Milián to send from there?

. . .

Where is it?

. . .

All right. You brought four?

. . .

Which ones are they?

. . .

That's fine, we are also receiving dwarf gandul beans which is the type we wish to try out. What is it? Dwarf?

. . .

Very likely. We'll try that. Where is it? Have you brought it here?

. . .

Listen, you harvested a great deal.

. . .

Good, thanks very much. Thank you, We'll see,

. . .

That's not very difficult. We can send you cartridges, but I'll let you in on the fact that 600 hunters, all excellent shots, are organizing to go to the rice plantations. [*Laughter*] What zone are you from? "From Aguada."

Ah, Aguada. Are there many rice plantations there?

"Yes."

And are there many wild ducks?

"Yes."

What do you need? Some cartridges there? Which do you prefer, cartridges or people to go and help hunt the wild ducks?

. . .

No, because they can not go now. I'll explain to you. These hunters are workers who hunt as a hobby, and they've taken on this hunt as volunteer work. Of course they are overjoyed.

Speak up, please.

. . .

Good. How many shells do you need there? How many? Speak "Thousands".

Thousands? [*Laughter*] How many wild ducks are in that area?

. . .

What local farmers' organization is that?

"Van Troi"

And does that include all the rice fields of that area?

. . .

Those who do not work their land will be given the opportunity to become true farmers or to sell their holdings

Good, that's fine. The day after tomorrow there will be a thousand shells in that rice plantation so you can shoot the wild ducks and prepare some wild duck fricassé, which is a real delicacy. [*Applause*].

There is one little problem comrades, that you have brought up here and that I have encountered frequently throughout the island. Three years ago I heard people grumbling about persons who have land and are not farmers, that is true.

I was in the province, and the Party, the workers, volunteer workers, farmers, everybody told me, "Well, there's a gentleman there whose cane has to be cut and his fields cleared every year; and he never comes near, he doesn't even supply the men with water, or aid of any sort. He doesn't lift a finger." This problem has been brought to our attention hundreds of times, because it is very annoying that we should ask workers to spend three months away from their families while they are cutting the cane on some landowner's farm—an absentee landlord living in town, with a business—perhaps a funeral home, another sugar plantation, a store or some other business—and let us not forget some members of the urban bourgeoisie went out to buy up land.

Anyone can throw up a roadside stand, run a jitney and thus get together 8 or 10 thousand pesos and buy a piece of land.

That is why we have announced that we will not recognize such purchases and ask your cooperation in not making illegal sales. We have explained clearly the policy that will be followed because there are all these problems now and later they will lead to additional problems.

The worker that is separated from his family for three months while away cutting cane, knows what resentment is felt in having to go there and cut cane of some man who never even goes there: That is unquestionable.

Well then, we feel that we must find a solution to this problem that is not too drastic. I don't think we should intervene the land, but give them the option of becoming real farmers, of moving to their farms and working the land. [*Applause*] If they don't want to be farmers, then they have the choice of selling the land. [*Applause*].

It seems to me that this is the best way to prevent problems fear, insecurity; give them the chance to become real farmers if they want to, and if they don't want to work the land, or can't for some reason or other,—to sell the land. And such cases should be discussed with the farmers' local committees to prevent problems of interventions of this type.

Does that seem fair enough to you? Do you consider it just? [*Applause*].

Well, comrades, I think the night has been characterized by my speaking of many things in a rather disorganized way. It was my fault, not yours. But I think that the main ideas, the main points of interest have been touched upon, and they were if a bit disorderly, at least I have tried to make a series of ideas as clear as possible so that all farmers will understand.

We have great confidence in the farmers; we feel that we understand the psychology of the farmer; we know his feeling of support for the Revolution, the farmers' loyalty to the Revolution. And that is the attitude that the Revolution will always maintain toward the farmers. The Revolution is educating the farmers' children; is making them technicians, agronomists, civil engineers, doctors, skilled workers. This process of the Revolution's social development, of educational development, will continue.

It is amazing how children have come to realize what things are important and to their interests

As I have already told you, we are going to send 100,000 young people to the Technological Institutes alone, between now and 1970. Many of them will be the sons and daughters of farmers. We are also going to increase the number of rural schools—in Las Villas Province, in Guane, in the Escambray Mountains—in a series of places we're going to open new schools, for teaching skills of various kinds. More than 80,000 students are graduating from sixth grade across the nation. And the young people are clear thinking; the youngsters are clear thinking.

It is amazing how children understand things that interest them. A few days ago we were in a valley in the south of the Sierra Maestra. On leaving—whenever we enter, some wait for us to come out—there were some children with papers

Among them was a small boy who gave me a letter. He lived nearby and his mother was there too. We had asked her something on the way into the valley. And the letter read: Major, we want your help, because we need a small parcel of land to produce crops, and a number of other things. The little boy had written the letter. And I asked the mother, "Where does your husband work?" He was working on the construction of the highway.

And I asked her: "Do you think that a parcel of land will solve any problems for either you or the country? Wouldn't it be better if we give the children scholarships?" And the children all shouted," "Yes, Yes".

And the boy with the letter also cried, "Yes, that is better." [*Laughter*].

And I said, "Look. I can't solve the problem of the land, because that isn't the policy we follow. Giving out small parcels of land doesn't solve anything. We are making ambitious plans precisely to solve the food problem; and that is not the way. But if you have many children, we can help." Immediately the rest appeared, all wanting scholarships. All the children, even the little brother, they all wanted the opportunity to study. And they understood that it was a thousand times better for them to have scholarships to study than to be given a small parcel of land on which they would be living in the same conditions they were already living in.

This is opening a new world to the children, is awakening their awareness. It is something impressive.

And we are thinking of continuing to develop plans in education. There is practically no part of the country that doesn't have schools now. And I believe that there are teachers in all corners of the country. There are some places where school facilities are very poor. The school is located in a bohío (thatched hut). We are also going to work on this and will continue improving communications in the rural areas. We will continue with the hospital program in the countryside. In other words, we can make even more rapid progress in the countryside than we have been making up to now.

In the next few years, production will increase considerably in all fields. Here we have talked about coffee, bananas, and milk. Speaking of milk, for example, the cross-breeding of dairy cattle with beef cattle that is, the crossing of Holstein and Zebu, is producing heifers that give 18, 19 and even 20 liters of milk per day at the age of 24 months, first calving.

We also expect that the second calving resulting from these crossings will produce as much as 25 liters of milk per day. Our milk problems will be solved. Next year 270,000 hectares of state land will be sown to pasture and 100,000 tons of fertilizer will be applied to pasturelands. You know the kind of care sugar cane is receiving now despite the acute drought. The months of March, April, and May have been rainless but we are not worried because the cane has received good care, including fertilizer. All sugar cane fields in the country—over 1,200,000 hectares of land—are now receiving 4 tons of fertilizer per *caballería*. All these canefields will receive at least two additional tons of ammonium nitrate, or three sprayings of foliar urea applied by airplane.

We are getting ready for a very large sugar harvest next year

Sixty planes are ready to handle the spraying. Even though it seems that we will have a dry year, we have so much faith in the results of good cultivation, the correct care of canefields, and fertilization, that we are confident that next year's sugar harvest will be considerably greater than this year's.

We do not want to cite any figures yet, but we are getting ready for a great harvest next year, principally through greater yield per **caballería**.

In the coming years, many of the problems we have been faced with will gradually disappear, problems such as supplies, the scarcity of **machetes**, wire, etc., as well as the problems of clothing and footwear. Many of these problems will gradually be eradicated. The growing of cotton is being increased and this year 1,000 **caballerías** will be planted to cotton and we plan to reach 3,000 **caballerías** next year.

Rice growing is also being increased considerably. In short, every item of agricultural production is receiving maximum stimulus from the Revolution.

Farm machinery and equipment such as bulldozers for roadbuilding, for water conservation projects and land-clearing, have been greatly increased in quantity; still greater resources will be made available to our small farmers.

In addition, I believe that we are now at the point at which the responsability for helping and handling supplies for the small farmers can be turned over in progressive stages to the rural aggregations, as you have urged. [*Applause*]

In the beginning, we considered these aggregations the strongest base upon which ANAP agricultural development could rest; but still two or three years ago, they lacked sufficient organization and control. Often they were not even able to meet their own goals, their own plans. And we feared that turning over the resources for developing ANAP agriculture, small farm agriculture, to the aggregations could lead to problems, that they would tend to carry out the plans of the State farms at the expense of the small farmers. That is not the situation today.

Now, the control, resources, and efficiency of the aggregations are much greater; we have told them that they must count on the ANAP, that it is not enough to take only State production into account. Often, they forgot about the other sector. How much cane? So much. And how much cane do the small farmers have? This is also part of the economy and it must be considered.

Now, many of our aggregation directors are truly interested in helping the small farmer and in raising small-farm production. In the future we shall require an accounting from the Secretary of the Party and the director of the aggregation, not only on State production but also on the aggregation of small farmers. [*Applause*]

And we shall not talk of how much cane has been produced by the aggregation, but rather how much has been produced by the region; not what quantity of root vegetables has been produced by the aggregation, but rather by the region. We are not going to do this suddenly, but in accordance with each regions possibilities, so as not to do things in a single day and then have trouble.

Now, for example, in the zone of Manzanillo, the agregation is going to help develop the rice production of all the small farmers of the region; that is, it is going to help them prepare the soil, solve all problems of irrigation, seeds, fertilizers, harvests, etc. In the region of Bayamo, the aggregation is going to help the small rice producers. It will also help the cattle raisers of the region in the planting of pasturage. I told them that next year we shall plant 1,000 **caballerías** there. In the region of Guane, the aggregation, through the Guane plan, is going to help the small farmers with tobacco, citrus-fruits, everything.

In other words, in accordance with the degree of organization and the resources of each aggregation, we are going to gradually turn responsibility over to them for the development of small-farm agriculture. We believe that we should not do this suddenly, from one day to the next, but over a period of from 6 to 8 months, region, by region, according to whether or not we are sure that the change will benefit agricultural development. Then all the fertilizer received will be sent to the different regions for distribution to both State agriculture and small farmers—wire, farm supplies, cord, financial resources, everything. And this will be a great simplification, because, at present, everything arrives via different routes: fertilizer comes one way, another product another way, etc, etc. We feel that the proposals you have made are very reasonable, and we are going to give them more dynamism so that everything will be much simpler. We believe that the separation, of supplies,—some from here and some from there—is definitely prejudicial.

In the next congress, comrades, I hope to have more time for discussion with all of you.

We have been trying during these last few years to learn, know and understand the problems of all of our agriculture; we have had many talks with farmers; we have been acquiring a great deal of knowledge. That is interesting and important for our work. We plan to keep on applying this method. We believe that there are many things still to be learned and there are many regions that we still don't know. Each time we visit the provinces, we try to tour a new region. We know practically all of them in Oriente, but we still don't know Velázquez, for example, where our most important bean producers are located; we still don't know some of Puerto Padre and Tunas, as well as some regions in Camagüey and Las Villas. But since we have always begun with the remotest and most mountainous regions, it will not be difficult to keep on getting to know and visiting the other regions of the country. These visits are always highly instructive.

We cannot win battles in agriculture if we do not know the terrain

And really, one of the most disagreeable facts for all of us to face is that we knew practically nothing about the geography of our country.

How can we even consider planting rice in the zones of Guamo, Birán and all those places, if we have never been there? How can many of our plans be carried out if we don't know the areas? Agriculture is very similar to war. In war one cannot win battles if he doesn't know the terrain, and in agriculture we cannot win battles if we don't know the terrain. We intend to acquire a deep knowlege of all the regions of the country so that the best regions can be chosen for all—projects, all plans, and all solutions.

We also believe that the small farmers must pay heed to technology, pay attention to methods of cultivation, to fertilization, irrigation and seed selection.

I assure you, comrades, that the present small-farm production can easily be doubled, and with still greater effort the present production of cane, coffee, tobacco, root vegetables and cattle can be tripled. We are developing a series of very concrete plans right here in the province of Havana and in other regions of the country in order to demonstrate to the small farmer that it is possible, through technology, to double and triple production with relative facility. Why? Because our present methods are very antiquated, our technology is very backward. And naturally, he who is accustomed, for example, to travel a road on a donkey, would be able to do it ten times faster by car—if the road were good—but he could go 30 times faster in an airplane. Of course, if a person changes the car for a plane he goes twice as fast; if he changes the donkey for a plane, he can go twenty times faster. And in agriculture, we are riding a donkey. That's why it isn't hard to double, triple, and in some cases quadruple and quintuple production, because we are traveling by donkey—no, worse!—we are traveling afoot. Afoot! And today, it is possible for us to take a plane in terms of technology. In other words, it is easy for us to multiply our production because we are starting from a technological zero; if we use a little technology, we can double and triple our production.

Our Revolution earnestly plans to reach these goals, not only in State agriculture, but in your agriculture—small farm agriculture. We would gain nothing by achieving enormous success in State agriculture, enormous progress in State agriculture, if we didn't achieve similar progress in your agriculture. In the interest of the nation, ANAP agriculture must develop parallel with State farm agriculture in technology. Some methods—that of the airplane, for example—cannot be used, but cultivation can be mechanized, tractors can be used, as well as fertilization, selected seeds, and a whole series of methods that, in spite of the disadvantages implicit in small parcels of land, can still be applied to a considerable extent.

This is what we expect from you, what we hope for from this Congress, which has been very good, very well organized—that, from now on, the small farmers resolve to apply technology, to technify their agriculture, to bring about a technical revolution in their agriculture, so that we do not have to feel the pain and the shame when we pass by those fields, those roads, of finding so much technological backwardness in our agriculture.

I know that the cane-growers, who have been growing cane all their lives—and I have seen that they are very enthusiastic about the proposals made here—are attached to cane; the tobacco-grower is attached to tobacco, and the coffee-grower is attached to his coffee. Each farmer loves the crop to which he is accustomed. Very well, let's show him how to cultivate that love, improving, increasing, doubling, tripling, quadrupling production. You all know perfectly the satisfaction one experiences when the cane is good, when it is big, when the coffee shrubs bear many beans, when the cows produce abundant milk.

It doesn't matter that work in the countryside is often hard. There are few spheres of activity that compensate man as does agriculture. There are few activities that compensate him and offer him as much satisfaction as does success in agriculture.

We should struggle against those who introduce vice and laziness into farm life

What we expect from you—naturally, with the Revolution's help, with maximum assistance on the part of the revolutionary government—is that you give a push to agricultural techniques, do rational things, have present the interests of the nation, act as genuine revolutionaries, deepen your awareness, censure speculators, struggle against those who introduce vices, who introduce laziness among the farmers. It is of greatest importance to deepen the revolutionary consciousness of the farmers, to abandon certain concepts that are not revolutionary, to forget about the price list, because really, those who go around with price lists to see what is to be planted, do not love agriculture, because he who loves cattle does not kill cows, rather, he does not starve cows to death to plant a carrot and make more money; he who loves his work, he who loves his crop, does not go around with a price list.

The price problems are getting solved. If a farmer from any sector thinks his crop is a lot work, that it doesn't pay enough, there is always a way to help him, to encourage him. to solve the problem. In any case, prices should be set for those crops where it is only fair that the price be improved, the most difficult crops, those that pay less, that are less encouraging for the farmer: but opportunistic prices should not be introduced to solve a problem with one product today and another problem tomorrow. That creates disorder.

That is, we expect a sense of responsability from the farmers, a sense of obligation to all the workers of the country, a sense of duty to the entire population.

The revolution, the country has done the maximum for the farmers. The workers of our country have done the maximum for our farmers. We, knowing that the farmers are revolutionary, knowing that the farmers have always helped the revolution with great enthusiasm, expect that in production, in work, they will have the same attitude towards the rest of country.

In the next congress I promise you that we are going to meet together for a longer time, I promise you that we are going to discuss many of those problems one by one. Everything that you have discussed is going to be analyzed, everything that can be solved is going to be solved; but in the next congress we are going to speak more of techniques, we are going to solve all those problems . . . the store, the tax, this, that and the other thing . . . all those problems, but in the next congress, which can be held in 1969—you have two years—the main topic will have to be techniques [*Applause*]. That is what is going to be discussed then: what methods are being applied, what are the most advanced ones, the best yields, the farmers that farm best, the vanguard farmers. But there have to be many more Mesas, many more Verdecias; that is, no one should be surprised at finding a farmer who is really progressive technologically, so that the most natural thing will be for all farmers to be progressive, to apply technology.

And in the next congress—I say this to the delegates—there should not be any representative of the farmers who does not have a good command of the problems of production technology. [*Applause*] Because the next congress in going to be an emminently technological one.

I am really very sorry, because I was very interested in this congress, that I did not have time to discuss many of the questions with you at greater length; but I, too, plan to prepare myself for the next congress, and I plan to have a lot more knowledge on all these problems; I plan to spend days discussing things with you, I plan to have more knowledge of the country's problems and of every crop. And sincerely, whenever I have time, I study. I recommend that the comrade farmers do the same thing. There are many farmers who know a great deal; and there are some who boast of knowing more than they really do. But in the next congress let them come prepared to discuss the technological problems of agriculture, let them come prepared, because it's going to be an examination. An examination! At the next congress we are going to ask all the farmers about the use of fertilizers, the formulas they use, what a formula is, what this means, what that means. You have to buckle down to study. Are we all in agreement? [*Shouts of "Yes"*].

By revolutionizing our agriculture, we shall be hitting the Yankees hard

So within two years—I won't say that we'll meet again then, since I intend to see you soon, anywhere, in any field, expect me there, because I'm planning to visit everywhere—within two years we are going to have a get-together here and discuss technological matters. And at the end of those two years, during which there will have been undoubted political advances, great advances in revolutionary awareness, we'll be interested in finding out just how much progress has been made in technological matters.

You told me, "Hit the Yankees hard." And we are going to hit the Yankees hard by revolutionizing our agriculture, overcoming our backward situation, applying technological know-how to our farming methods. The hardest blow the Yankees are going to get is when they realize that we are producing 10,000,000 tons of sugar [*Applause*], when it dawns on them that despite their blockade, despite their aggressions, we have completely overcome our difficulties. We will do away with rationing, and solve our food supply problems, and not at the level of nutrition before, not at that standard. This blow, one of the hardest that the Yankees are going to receive, will come about when we have—despite all the obstacles they have attempted to put in our way—within a few years' time, by 1970, eliminated ration books and all such problems. And with a consumption level per capita twice as high as any known by our people before the Revolution. [*Applause*] Of course, we are not working so hard in economic matters with the sole or fundamental motive of obtaining a higher standard of living, of getting richer, no. For us the struggle for advancement of the economy, for economic successes, forms part of our ideology.

We have our ideas about how socialism and communism should be built, and the results are what will prove who is right. And this does not just apply to our capitalist adversaries those who want the socialist revolution to fail; within the socialist camp itself there are differing concepts of how socialism and communism should be built. And while we are developing our people's revolutionary awareness while we are deepening our ideology and internationalist awareness, we desire and intend to show that in the field of economic development, and in the field of building socialism and communism the way we have selected is the most correct, the most revolutionary way.

    Patria o muerte!
    Venceremos!
    [*Ovation*]

COMMUNIST CONTINUITY AND PROGRAM

Labor, Nature, and the Evolution of Humanity
The Long View of History
FREDERICK ENGELS, KARL MARX
GEORGE NOVACK
MARY-ALICE WATERS

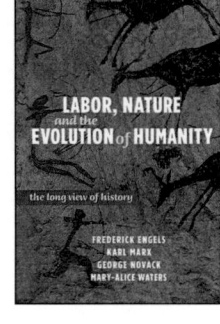

Without understanding that social labor, transforming nature, has driven humanity's evolution for millions of years, working people are unable to see beyond the capitalist epoch of class exploitation that warps all human relations, ideas, and values. Only the revolutionary conquest of state power by the working class can open the door to a world free of capitalist exploitation, degradation of nature, subjugation of women, racism, and war. A world built on human solidarity. A socialist world. $12. Also in Spanish and French.

The Fight for a Workers and Farmers Government in the United States
JACK BARNES

The shared exploitation of workers and working farmers by banking, industrial, and commercial capital lays the basis for their alliance in a revolutionary fight for a government of the producers. In *New International* no. 4. $14

The Workers and Farmers Government
JOSEPH HANSEN

How experiences in post–World War II revolutions in Yugoslavia, China, Algeria, and Cuba enriched communists' theoretical and practical understanding of revolutionary governments of the workers and farmers. "What is involved is governmental power," writes Hansen, "the possibility of smashing the old structure and overturning capitalism." $5

The Organizational Character of the Socialist Workers Party
1965 Resolution of the SWP

Deepening capitalist crisis and sharpening class conflict demand a revolutionary solution. Active preparation for such struggles determines the kind of organization the Socialist Workers Party has set out to build from its birth. $5. Also in Spanish.

The Low Point of Labor Resistance Is Behind Us
The Socialist Workers Party Looks Forward
JACK BARNES, MARY-ALICE WATERS
STEVE CLARK

The global order imposed by victors of the inter-imperialist slaughter of World War II is shattering, with explosive ramifications for workers and farmers worldwide. A long retreat by the working class and unions has come to an end. More and more workers of all ages, skin colors, and both sexes are saying, "Enough is enough!" This book highlights opportunities ahead for class-conscious workers to forge a labor party built on the unions. And a mass proletarian vanguard able to lead the struggle to end capitalist rule, opening a future for humanity. $10. Also in Spanish and French.

Revolutionary Continuity
Marxist Leadership in the U.S.
The Early Years, 1848–1917
Birth of the Communist Movement, 1918–1922
FARRELL DOBBS

"Successive generations of proletarian revolutionists have participated in the movements of the working class and its allies. . . . Marxists today owe them not only homage for their deeds. We also have a duty to learn what they did wrong as well as right so their errors are not repeated." —*Farrell Dobbs*. Two volumes, $17 each.

The Transitional Program for Socialist Revolution
LEON TROTSKY

The Socialist Workers Party program, drafted by Trotsky in 1938, still guides the SWP and communists the world over. The party "uncompromisingly gives battle to all political groupings tied to the apron strings of the bourgeoisie. Its task—the abolition of capitalism's domination. Its aim—socialism. Its method—the proletarian revolution." $17. Also in Farsi.

The Clintons' Anti-Working-Class Record
Why Washington Fears Working People
JACK BARNES

What working people need to know about the profit-driven course of Democrats and Republicans alike over the last three decades. And the political awakening of workers seeking to understand and resist the capitalist rulers' assaults. $10. Also in Spanish, French, Farsi, and Greek.

Cuba and the Coming American Revolution
JACK BARNES

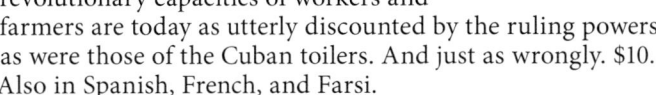

This is a book about the struggles of working people in the imperialist heartland, the youth attracted to them, and the example set by the Cuban people that revolution is not only necessary—it can be made. It is about the class struggle in the US, where the revolutionary capacities of workers and farmers are today as utterly discounted by the ruling powers as were those of the Cuban toilers. And just as wrongly. $10. Also in Spanish, French, and Farsi.

Opening Guns of World War III: Washington's Assault on Iraq
JACK BARNES

The murderous assault on Iraq in 1990–91 heralded increasingly sharp conflicts among imperialist powers, growing instability of capitalism, and more wars. Also includes:

1945: When US Troops Said No! by Mary-Alice Waters

Lessons from the Iran-Iraq War by Samad Sharif

In *New International* no. 7. $14. Also in Spanish, French, and Farsi.

The Revolution Betrayed
What Is the Soviet Union and Where Is It Going?
LEON TROTSKY

In 1917 workers and peasants of Russia were the motor force for one of the deepest revolutions in history. Yet within ten years a political counterrevolution by a privileged social layer, whose chief spokesperson was Joseph Stalin, was being consolidated. The classic study of the Soviet workers state and its degeneration. $17. Also in Spanish, Farsi, and Greek.

Colombia: Fidel Castro on the Debate around Revolutionary Strategy and Lessons of the Cuban Revolution
FROM THE PAGES OF THE *MILITANT*

Excerpts from Fidel Castro's *Peace in Colombia* and articles from the *Militant*. In describing the Cuban leadership's efforts to end decades of war between the FARC guerrilla movement and Colombia's brutal regime, Castro in his prologue, afterword, and other statements explains why Cuban revolutionaries, unlike FARC leaders, rejected taking hostages and organized working people to win state power, not pursue a "prolonged people's war." $5. Also in Spanish.

In Defense of the US Working Class
MARY-ALICE WATERS

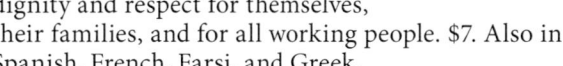

Drawing on the fighting traditions of the oppressed and exploited of all colors and national origins, in 2018 tens of thousands of teachers and other working people in West Virginia, Oklahoma, and other states waged victorious strikes. They fought for dignity and respect for themselves, their families, and for all working people. $7. Also in Spanish, French, Farsi, and Greek.

Lenin's Final Fight
Speeches and Writings, 1922–23
V.I. LENIN

In 1922 and 1923, V.I. Lenin, central leader of the world's first socialist revolution, waged what was to be his last political battle—one that was lost following his death. At stake was whether that revolution, and the international communist movement it led, would remain on the revolutionary proletarian course that brought workers and peasants to power in October 1917. $17. Also in Spanish, Farsi, and Greek.

The Struggle for a Proletarian Party
JAMES P. CANNON

"The workers of America have power enough to topple the structure of capitalism at home and to lift the whole world with them when they rise," Cannon asserts. On the eve of World War II, a founder of the communist movement in the US and leader of the Communist International in Lenin's time defends the program and party-building norms of Bolshevism. $20. Also in Spanish and Farsi.

WWW.PATHFINDERPRESS.COM

Also from Pathfinder

The Communist Manifesto
Karl Marx and Frederick Engels

Communism, say the founding leaders of the revolutionary workers movement, is not a set of ideas or preconceived "principles" but workers' line of march to power, springing from a "movement going on under our very eyes." $5. Also in Spanish, French, Farsi, and Arabic.

Socialism: Utopian and Scientific
Frederick Engels

"To make men the masters of their own form of social organization—to make them free—is the mission of the modern proletariat," writes Engels. A classic guide to the operations of capitalism and struggles of the working class. $10. Also in Farsi.

The Teamster Series
Farrell Dobbs

Four books on the strikes, organizing drives, and political campaigns that transformed the Teamsters across the Midwest in the 1930s into a militant industrial union movement. Written by Farrell Dobbs, the general organizer of these Teamster battles and leader of the Socialist Workers Party.

A tool for workers seeking to use union power in every workplace and advance the fight for an independent labor party. $16 each, series $50. Also in Spanish. *Teamster Rebellion* is also available in French, Farsi, and Greek.

Are They Rich Because They're Smart?
CLASS, PRIVILEGE, AND LEARNING UNDER CAPITALISM
Jack Barnes

Exposes growing class inequalities in the US and the self-serving rationalizations of well-paid professionals who think their "brilliance" equips them to "regulate" working people, who don't know what's in our own best interest. $10. Also in Spanish, French, Farsi, and Arabic.

New International
A MAGAZINE OF MARXIST POLITICS AND THEORY

CAPITALISM'S LONG HOT WINTER HAS BEGUN
JACK BARNES

Today's global capitalist crisis is but the opening stage of decades of economic, financial, and social convulsions and class battles. Class-conscious workers confront this historic turning point for imperialism with confidence, Jack Barnes writes, drawing satisfaction from being "in their face" as we chart a revolutionary course to take power. In *New International* no. 12. $14. Also in Spanish, French, Farsi, Arabic, and Greek.

OUR POLITICS START WITH THE WORLD
JACK BARNES

The huge economic and cultural inequalities between imperialist and semicolonial countries, and among classes within them, are accentuated by the workings of capitalism. To build parties able to lead a successful revolutionary struggle for power in our own countries, vanguard workers must be guided by a strategy to close this gap. In *New International* no. 13. $14. Also in Spanish, French, Farsi, and Greek.

U.S. IMPERIALISM HAS LOST THE COLD WAR
JACK BARNES

The collapse of regimes across Eastern Europe and the USSR claiming to be communist did not mean workers and farmers there had been crushed. In today's sharpening capitalist conflicts and wars, these toilers are joining working people the world over in the class struggle against exploitation. In *New International* no. 11. $14. Also in Spanish, French, Farsi, and Greek.

Cointelpro
THE FBI'S SECRET WAR ON POLITICAL FREEDOM
Nelson Blackstock

An in-depth look at the 1960s and '70s covert FBI disruption and counterintelligence program—code-named COINTELPRO. Contains reproductions of FBI documents released through the Socialist Workers Party suit against government spying. $15

"It's the Poor Who Face the Savagery of the US 'Justice' System"
THE CUBAN FIVE TALK ABOUT THEIR LIVES WITHIN THE US WORKING CLASS

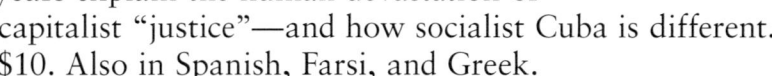

How US cops, courts, and prisons work as "an enormous machine for grinding people up." Five Cuban revolutionaries framed up and held in US jails for 16 years explain the human devastation of capitalist "justice"—and how socialist Cuba is different. $10. Also in Spanish, Farsi, and Greek.

The Jewish Question
A MARXIST INTERPRETATION
Abram Leon

Why is Jew-hatred still raising its ugly head? What are its class roots—from antiquity through feudalism, to capitalism's rise and current crises? Why is there no solution under capitalism? The author, Abram Leon, was killed in the Nazi gas chambers. Revised translation, new introduction, and 40 pages of illustrations and maps. $17. Also in Spanish and French.

Women's Liberation and the African Freedom Struggle
Thomas Sankara

"There is no true social revolution without the liberation of women," explains the leader of the 1983–87 revolution in the West African country of Burkina Faso. $5. Also in Spanish, French, and Farsi.

WWW.PATHFINDERPRESS.COM

THE CUBAN REVOLUTION AND WORLD POLITICS

October 1962
The 'Missile' Crisis as Seen from Cuba
TOMÁS DIEZ ACOSTA

In October 1962 Washington pushed the world to the edge of nuclear war. Here the full story of that historic moment is told from the perspective of the Cuban people, whose determination to defend their sovereignty and their socialist revolution blocked US plans for a devastating military assault. $17

Red Zone
Cuba and the Battle against Ebola in West Africa
ENRIQUE UBIETA GÓMEZ

When three African countries were hit in 2014–15 by the Ebola epidemic, Cuba's revolutionary government sent what no other country even pretended to provide: more than 250 volunteer doctors, nurses, and other medical workers. This firsthand account of their actions shows the kind of men and women only a socialist revolution can produce. $17. Also in Spanish and French.

The First and Second Declarations of Havana

Nowhere are the questions of revolutionary strategy that today confront men and women on the front lines of struggles in the Americas addressed with greater truthfulness and clarity than in these uncompromising indictments of imperialist plunder and "the exploitation of man by man." Adopted by million-strong assemblies of the Cuban people in 1960 and 1962. $10. Also in Spanish, French, Farsi, Arabic, and Greek.

'Cuba Will Never Adopt Capitalist Methods'
FIDEL CASTRO

Cuba's rectification process, its contributions to building socialism worldwide, and the victory of Cuban-Angolan-SWAPO forces against the South African army in southern Angola in early 1988. $5

Our History Is Still Being Written
The Story of Three Chinese Cuban Generals in the Cuban Revolution
ARMANDO CHOY, GUSTAVO CHUI
MOISÉS SÍO WONG
MARY-ALICE WATERS

"What was the key measure to uproot discrimination against Chinese and blacks in Cuba? It was the socialist revolution itself." New edition sheds light on Chinese Cubans' involvement in Cuba's internationalist course, including in Africa and Latin America. $15. Also in Spanish, French, Farsi, Greek, and Chinese.

Women in Cuba: The Making of a Revolution within the Revolution
VILMA ESPÍN, ASELA DE LOS SANTOS
YOLANDA FERRER

The integration of women in the ranks and leadership of the Cuban Revolution was intertwined with the proletarian course of the leadership of the revolution from the start. This is the story of that revolution and how it transformed the women and men who made it. $17. Also in Spanish, Farsi, and Greek.

Cuba and Angola: The War for Freedom
HARRY VILLEGAS ("POMBO")

The story of Cuba's unparalleled contribution to the fight to free Africa from the scourge of apartheid. And how, in the doing, Cuba's socialist revolution was strengthened. $10. Also in Spanish, Farsi, and Greek.

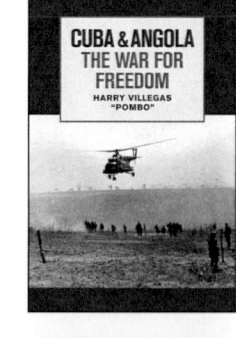

Che Guevara Talks to Young People

Guevara challenges the youth of Cuba and the world to work. To become disciplined. To join the vanguard on the front lines of struggles, small and large. To become a different kind of human being as they fight together with working people of all lands to transform the world. $12. Also in Spanish and Greek.

BUILDING A PROLETARIAN PARTY

In Defense of Marxism
Against the Petty-Bourgeois Opposition in the Socialist Workers Party
LEON TROTSKY

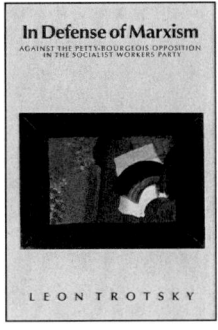

A reply to those in the revolutionary workers movement in the late 1930s bending to bourgeois patriotism during Washington's buildup to enter World War II. Trotsky explains why only a party fighting to bring workers into its ranks and leadership can steer a communist course. In the process, he defends the materialist and dialectical foundations of Marxism. $17. Also in Spanish.

Is Socialist Revolution in the US Possible?
A Necessary Debate among Working People
MARY-ALICE WATERS

An unhesitating "Yes"—that's the answer given here. Possible—but not inevitable. That depends on what working people do. $7. Also in Spanish, French, and Farsi.

Capitalism's World Disorder
Working-Class Politics at the Millennium
JACK BARNES

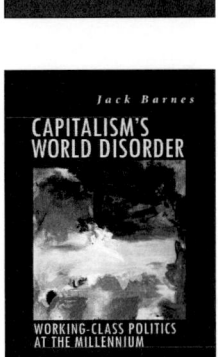

The social devastation and financial crises, the coarsening of politics, the cop brutality and acts of imperialist aggression accelerating around us—all are products not of something gone wrong with capitalism but of its lawful workings. Yet the future can be changed by the united struggle and selfless action of working people conscious of their power to transform the world. $20. Also in Spanish and French.

What Is to Be Done?
V.I. LENIN

The stakes in creating a disciplined organization of working-class revolutionaries capable of acting as a "tribune of the people, able to react to every manifestation of tyranny and oppression, no matter where it appears, to clarify for all and everyone the world-historic significance of the struggle for the emancipation of the proletariat." Written in 1902. $20

Socialism on Trial
Testimony at Minneapolis Sedition Trial
JAMES P. CANNON

The revolutionary program of the working class, presented in response to frame-up charges of "seditious conspiracy" in 1941, on the eve of US entry into World War II. The defendants were leaders of the Minneapolis labor movement and the Socialist Workers Party. $15. Also in Spanish, French, and Farsi.

The Emancipation of Women
V.I. LENIN

Women's emancipation, Lenin wrote, begins "only when an all-out struggle begins, led by the proletariat wielding state power," to draw women as equals into productive social labor. And as cooking, childcare, and other housework are transformed into social tasks of "a large-scale socialist economy." $7

Their Trotsky and Ours
JACK BARNES

To lead the working class in a successful revolution, a mass proletarian party is needed whose cadres, well beforehand, have absorbed a world communist program, are proletarian in life and work, derive deep satisfaction from doing politics, and have forged a leadership with an acute sense of what to do next. This book is about building such a party. $12. Also in Spanish, French, and Farsi.

The History of American Trotskyism, 1928–38
Report of a Participant
JAMES P. CANNON

"Trotskyism is not a new movement, a new doctrine," Cannon says, "but the restoration, the revival of genuine Marxism as it was expounded and practiced in the Russian Revolution and in the early days of the Communist International." Talks by a founding leader of American communism on building a proletarian party in the United States. $17. Also in Spanish and French.

WWW.PATHFINDERPRESS.COM

EXPAND YOUR REVOLUTIONARY LIBRARY

Thomas Sankara Speaks
The Burkina Faso Revolution, 1983–87

Under Sankara's guidance, Burkina Faso's revolutionary government led peasants, workers, women, and youth to expand literacy; to sink wells, plant trees, erect housing; to combat women's oppression; to carry out land reform; to join others worldwide to free themselves from the imperialist yoke. $20. Also in French.

Imperialism's March toward Fascism and War
JACK BARNES

"There will be new Hitlers, new Mussolinis. That is inevitable. What is not inevitable is that they will triumph. The working-class vanguard will organize our class to fight back against the devastating toll we are made to pay for the capitalist crisis. The future of humanity will be decided in the contest between these contending class forces."
In *New International* no. 10. $14. Also in Spanish, French, Farsi, and Greek.

By Any Means Necessary
MALCOLM X

"The imperialists know the only way you will voluntarily turn to the fox is to show you a wolf." In eleven speeches and interviews, Malcolm X presents a revolutionary alternative to this reformist trap, taking up political alliances, women's rights, US intervention in the Congo and Vietnam, capitalism and socialism, and more. $15

Democracy and Revolution
GEORGE NOVACK

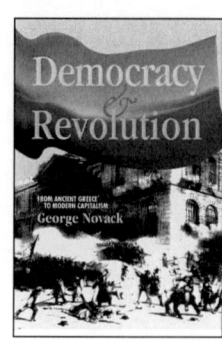

The limitations and advances of various forms of democracy in class society, from its roots in ancient Greece through its rise and decline under capitalism. Discusses the emergence of Bonapartism, military dictatorship, and fascism, and how democracy will be advanced under a workers and farmers regime. $17

Puerto Rico: Independence Is a Necessity
RAFAEL CANCEL MIRANDA

One of the five Puerto Rican Nationalists imprisoned by Washington for more than 25 years and released in 1979 speaks out on the brutal reality of US colonial domination, the example of Cuba's socialist revolution, and the ongoing struggle for independence. $5. Also in Spanish and Farsi.

Capitalism and the Transformation of Africa
Reports from Equatorial Guinea
MARY-ALICE WATERS
MARTÍN KOPPEL

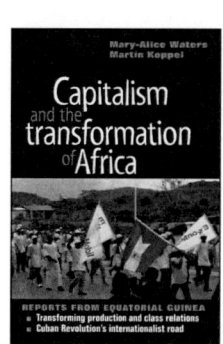

Describes how, as Equatorial Guinea is pulled into the world market, both a capitalist class and a working class are being born. Also documents the work of volunteer Cuban health-care workers there—an expression of the living example of Cuba's socialist revolution. $10. Also in Spanish and Farsi.

Pathfinder Press accessible ebooks for the blind, those with low vision, or other challenges reading print books

For a list of current accessible titles, go to: pathfinderpress.com/collections/books-for-the-blind.

Visit Bookshare.org for information on how to sign up.